Information Technology and Global Governance

Series Editor
Derrick L. Cogburn
American University
Bethesda, MD, USA

Information Technology and Global Governance focuses on the complex interrelationships between the social, political, and economic processes of global governance that occur at national, regional, and international levels. These processes are influenced by the rapid and ongoing developments in information and communication technologies. At the same time, they affect numerous areas, create new opportunities and mechanisms for participation in global governance processes, and influence how governance is studied. Books in this series examine these relationships and influences.

Edoardo Celeste • Nicola Palladino
Dennis Redeker • Kinfe Yilma

The Content Governance Dilemma

Digital Constitutionalism, Social Media
and the Search for a Global Standard

palgrave
macmillan

Edoardo Celeste
Dublin City University
Dublin, Ireland

Nicola Palladino
Trinity College Dublin
Dublin, Ireland

Dennis Redeker
University of Bremen
Bremen, Germany

Kinfe Yilma
Addis Ababa University
Addis Ababa, Ethiopia

Information Technology and Global Governance
ISBN 978-3-031-32923-4   ISBN 978-3-031-32924-1   (eBook)
https://doi.org/10.1007/978-3-031-32924-1

Cover illustration: Pattern © Melisa Hasan

This Palgrave Macmillan imprint is published by the registered company Springer Nature Switzerland AG.
The registered company address is: Gewerbestrasse 11, 6330 Cham, Switzerland

ACKNOWLEDGEMENTS

The initial idea for this research project was developed in December 2019 during a meeting of the "Digital Constitutionalism Network" hosted and funded by the Center for Advanced Internet Studies (CAIS) in Bochum, Germany. This was our last in-person meeting before the pandemic, and, still unaware of the challenges that we would have faced a few months later, we successfully applied for a grant funded by Facebook Research. Our project, entitled 'Digital Constitutionalism: In Search of a Content Governance Standard', allowed us to fund part of our research activities during the pandemic and also kept us in contact—a weekly online meeting that we all looked forward to!

For this, we are very grateful to each other, for the company, mutual patience, perseverance and motivation given to each other. The present book is the result of the collaboration among the four of us over the past three years. Each of us contributed equally in the writing process and our names are listed in alphabetical order. However, this book would not have been possible without the support of Priya Agarwal, Selim Başoğlu, Mariana Bernardes da Silva Passos, Justinas Kuprys, Cerys Lee and Sorcha Montgomery, who helped us with the data collection and multimedia aspects of the project.

We would like to thank all our colleagues from the Digital Constitutionalism Network with whom we have discussed the core ideas at the basis of this project, and in particular Clara Iglesias Keller and Amélie Heldt for reviewing an earlier paper that condensed the main arguments of this book. We presented preliminary findings of our work at a webinar entitled 'Digital Constitutionalism and Content Governance:

Social Media After the Capitol Hill Events', and we would like to thank the panellists—Prof Maura Conway, Prof Eugenia Siapera, Dr Tanja Lokot, Dr Suzanne Little, Dr Rishi Gulati and Dr Anne Marieke Mooij—for helping us contextualise our work in the socio-political situation we were living. We also thank the attendees of the AoIR Conference 2022 in Dublin for their valuable questions and comments when we presented the latest draft of this work.

Kinfe Yilma has received various forms of support from colleagues at Addis Ababa University, including the Office of the Vice President for Research and Technology Transfer and the School of Law. Nicola Palladino has received funding from the European Union's Horizon 2020 Research and Innovation Programme under the Human+ Cofund Marie Skłodowska-Curie grant agreement No. 945447. We are all extremely grateful to our colleagues, administrative staff, research support officers, heads and deans for helping us complete this project during very challenging times.

On a personal note, we wish to recognise the unwavering support of our families, friends and partners over the past few years. Their encouragement, backing and reassurance gave us the reason to persevere in this project. Finally, we would like to thank the editorial team at Palgrave, the series editor and the anonymous reviewers for their patience and assistance as well as the Center for Advanced Internet Studies (CAIS) in Bochum, Germany, which allowed us to publish this book in Open Access format.

CONTENTS

About the Authors

Edoardo Celeste (PhD in Law) is Assistant Professor of Law, Technology and Innovation at the School of Law and Government of Dublin City University, Ireland. Edoardo is the Programme Chair of the Erasmus Mundus Master in Law, Data and AI (EMILDAI), the coordinator of the DCU Law and Tech Research Cluster and the Deputy-Director of the DCU Law Research Centre. His research interests lie in the field of digital rights and constitutionalism, privacy and data protection law, and online platforms governance and regulation. He is the author of *Digital Constitutionalism: The Role of Internet Bills of Rights* (2022) and the editor of *Data Protection Beyond Borders* (2021), *Constitutionalising Social Media* (2022) and *Data Protection and Digital Sovereignty Post-Brexit* (2023). He is a founding member of the Digital Constitutionalism Network.

Nicola Palladino (PhD in Sociology, Social Analysis and Public Policy) is a research fellow under the Human+ Cofund Marie Skłodowska-Curie programme at the Trinity Long Room Hub Arts and Humanities Research Institute, Trinity College Dublin, Ireland. He is also a member of the Digital Constitutionalism Network and of the Internet & Communication Policy Center at the University of Salerno. His main research interests include global Internet governance processes, digital constitutionalism, platform governance and content governance and AI ethics and regulation. He recently published the volume *Legitimacy, Power, and Inequalities* *in the Multistakeholder Internet Governance: Analyzing IANA Transition* within the Palgrave Information Technology and Global Governance book series.

Dennis Redeker (PhD in Political Science) is a postdoctoral researcher at the University of Bremen's ZeMKI, Centre for Media, Communication and Information Research. Dennis is also a fellow at the Information Society Law Center at the University of Milan (Statale), a young academic fellow at the Academy of Sciences in Hamburg and a visiting professor at the Center for Technology and Society (CTS) at FGV Direito Rio (Rio de Janeiro). He is a founding member of the Digital Constitutionalism Network. Dennis' research interests include the legitimacy of content governance on social media platforms, the governance and geopolitics of emerging technologies, and the role of digital rights advocacy networks in shaping Internet governance.

Kinfe Yilma (PhD in Law and Technology) is Assistant Professor of Law at Addis Ababa University School of Law. Kinfe's research interests lie at the intersections of technology law, human rights, digital constitutionalism and law reform. He has published extensively in these fields, including more recently a monograph titled *Privacy and the Role of International Law in the Digital Age* (2023). Kinfe is a founding member of the Digital Constitutionalism Network and the Platform Governance Research Network.

ABBREVIATIONS

AI	Artificial Intelligence
CCPR	Human Rights Committee
CDA	Communications Decency Act
CERD	International Convention on the Elimination of All Forms of Racial Discrimination
CoE	Council of Europe
DMCA	Digital Millennium Copyright Act
DSA	Digital Services Act
ECHR	European Convention on Human Rights and Fundamental Freedoms
EU	European Union
GFMD	Global Forum for Media Development
GNI	Global Network Initiative
HR Committee	Human Rights Committee
HRC	Human Rights Council
IBRs	Internet Bills of Rights
ICCPR	International Covenant on Civil and Political Rights
ICESCR	International Covenant on Economic, Social and Cultural Rights
MAU	Monthly Active Users
NetzDG	Network Enforcement Act
NGOs	Non-governmental Organisations
OHCHR	Office of the High Commissioner for Human Rights
PATA	Platform Accountability and Transparency Act
PECA	Pakistan's Electronic Crimes Act
UDHR	Universal Declaration of Human Rights
UN	United Nations

UNESCO	The United Nations Educational, Scientific and Cultural Organization
UNGA	The United Nations General Assembly
UNGPs	United Nations Guiding Principles on Business and Human Rights
US	United States

LIST OF FIGURES

LIST OF TABLES

CHAPTER 1

Introduction

Abstract One of the main issues of global social media governance relates to the definition of the rules governing online content moderation worldwide. One could think that it would be sufficient for online platforms to refer to existing international human rights standards. However, a more careful analysis shows that international law provides exclusively general principles and that a single human rights standard does not exist. Since their inception, major social media platforms have set their own rules; yet this normative autonomy too has raised serious concerns. The current situation exposes a dilemma for online content governance that seriously affects the operations of social media companies and impacts on the exercise of fundamental rights by users as well as digital policy strategies.

Keywords Online content governance • Social media • Normative dilemma • Private norms • International law • Constitutionalisation • Digital constitutionalism

One of the main issues of global social media governance relates to the definition of the rules governing online content moderation worldwide. One could think that it would be sufficient for online platforms to refer to existing international human rights standards. However, a more careful analysis shows not only that international law provides exclusively general principles, which do not specifically address the context of online content

© The Author(s) 2023
E. Celeste et al., *The Content Governance Dilemma*, Information
Technology and Global Governance,
https://doi.org/10.1007/978-3-031-32924-1_1

moderation. But also that a single human rights standard does not exist, as even the same provisions and principles are interpreted by courts in different ways across the world. This is one of the reasons why, since their inception, major social media platforms have set their own rules, adopting their own peculiar language, values and parameters. Yet, at the same time, this normative autonomy too has raised serious concerns. Why should private companies establish the rules governing free speech online? Is it legitimate to depart from minimal human rights standards and impose more (or less) stringent rules?

The current situation exposes a dilemma for online content governance that seriously affects the operations of social media companies and impacts on the exercise of fundamental rights by users as well as digital policy strategies. On the one hand, if social media platforms simply adopted international law standards, they would be compelled to operate a choice on which interpretative model to follow—for example, between a US-style freedom of expression-dominated approach and a European-style standard, which tries to balance freedom of expression with other social values. And the same would be if they decided to adopt the law of one country. Moreover, they would also need to put in place a mechanism able to translate, or 'operationalise', such general state-centred standards in the context of online content moderation. On the other hand, where social media platforms adopt their own values, rules and terminology to regulate content moderation, thus departing from international law standards, they are accused of censorship or laxity, intrusiveness or negligence.

The present work aims to analyse this normative dilemma. Chapter 2, entitled 'The Content Governance Dilemma', sets the scene, deconstructing the core elements of the conundrum. We analyse the evolution from community-based and user-led online content moderation to the professionalised, industry-led sets of mechanisms that characterise it today. Such a transition, combined with the intervention of external actors in defining rules for online content, such as national legislators and courts, determined the emergence of a macro governance dimension of content moderation. Its conception of the social media environment as a public forum clashes with the private approach of its micro dimension, which is completely dominated by the platforms themselves and managed as a private independent space. This tension lies at the basis of a complex normative dilemma, whose central question is: Which rules should govern content online? Private norms or democratically voted laws? If more national laws or international standards are simultaneously applicable, which law is to be chosen? How does one avoid the risk of having one-single approach imperialistically imposed on the others? In this chapter, we explain how this dilemma exposes a tension between the risk of normative authoritarianism, anomie and imperialism. A process of 'constitutionalisation' of the social media environment seems to be needed in any case. Social media internal rules should better incorporate fundamental rights and guarantees. An input to this process is increasingly originating from civil society actors. Over the past few years, a significant number of 'bills of rights' have been proposed to articulate constitutional rights for social media. This phenomenon has been described in terms of the emergence of a movement of 'digital constitutionalism'. In this book, after having analysed the contribution that international human rights law can give to the constitutionalisation of social media content moderation standards, we focus our analysis on the role and message of civil society initiatives in this field.

Vis-à-vis the twofold issue of constitutional authoritarianism of online platforms' terms of service and the potential normative imperialism of imposing one dominant legal approach, many have advocated the application of international human rights standards as a solution to the issues of platform governance. Chapter 3, entitled 'The International Law of Content Governance', examines the question of whether, and the extent to which, international law really offers normative guidance to the complex world of platform content governance. It argues that the potential of international human rights law in offering much-needed normative guidance to content governance is circumscribed by three interrelated factors. First, international human rights law is—by design—state-centred and hence does not go a long way in attending to human rights concerns in the private sector. Second, international human rights law standards are couched in general principles, and hence, less suited to apply in the context of platform content moderation which requires a rather granular and dynamic system of norms. Third, and related to the second, the generic international content governance standards have not adequately been unpacked by relevant adjudicative bodies to make them fit for purpose to the present the realities of content moderation. The chapter then maps applicable content governance standards in international law, focusing in particular on the role of soft law instruments addressing private organisations.

Chapter 4, entitled 'Shaping Standards from Below: Insights from Civil Society', proposes the analysis of civil society impulses in the field of online content moderation, a source that contributes to the definition of normative standards that has been so far neglected by the scholarship. Internet bills of rights promoted by civil society are presented as expressing the 'voice' of communities that struggle to propose an innovative message within traditional institutional channels: one of the layers of the complex process of constitutionalisation that is pushing towards reconceptualising core constitutional principles in light of the challenges of the digital society in a new form of 'digital constitutionalism'. This chapter illustrates the findings of a content analysis of 40 Internet bills of rights that include principles related to online content governance. We start with an overview of the main features of the textual corpus, taking into account the distribution across time and geographical areas of these bottom-up sources of constitutional values. We illustrate the main principles, rights and standards detected in the corpus and their mutual relationships, tracing back a civil society framework for content moderation. Our analysis then focuses on the substantive standards promoted by these declarations and on the procedural guarantees articulating formal rules and procedures through which substantive rights are created, exercised and enforced. We then analyse some ad hoc provisions specifically crafted to address social media platforms, which most of the times are attempts to contextualise and adapt international human rights standards into more granular norms and rules to be implemented in the platform environment.

Chapter 5, entitled 'Platform Policies Versus Human Rights Standards', investigates to what extent human rights standards as enshrined in international law and in civil society initiatives are reflected in social media platforms' standards—both on paper and in terms of adopted practices. The chapter utilises a comparative approach and studies five major social media platforms—Facebook and Instagram (Meta Inc.), TikTok/Douyin (Bytedance Inc.), Twitter (Twitter Inc.) and YouTube (Alphabet Inc.). This chapter first explores the official commitments of the three platforms to international human rights. Building on data provided by the platforms themselves, it then compares the magnitude of cases of moderation occurring between different principles and across platforms for 2021, the last full available year. While a number of interesting differences and commonalities exist between the platforms in terms of substantive content moderation outcomes, an investigation of procedural practices is as important to understand how platforms moderate. Hence, the chapter subsequently compares procedural principles demanded by civil society groups and those that can be derived from international human rights law with platform practices, specifically concerning transparency reporting and automated content moderation. The chapter finds a relatively high degree of convergence among the platforms on a number of practices.

Chapter 6 concludes the book by highlighting the role played by civil society actors in the broader process of constitutionalisation of social media internal rules. We argue that the solution to the content governance dilemma lies in its composite nature. No actor has the final word, but we rather witness a polyphonic conversation. Multiple societal layers are simultaneously and gradually contributing to rearticulate the core principles of contemporary constitutionalism in the context of online content moderation.

CHAPTER 2

The Content Governance Dilemma

Abstract Social media content moderation faces a complex normative dilemma. The central question is: Which rules should govern online content? Private norms or democratically voted laws? If more national laws or international standards are simultaneously applicable, which law to choose? How to avoid the risk of having one single approach imperialistically imposed on the others? This dilemma exposes a tension between the risk of normative authoritarianism, imperialism and anomie. A process of 'constitutionalisation' of the social media environment seems to be needed. An input to this process is increasingly originating from civil society actors. Over the past few years, a significant number of 'bills of rights' have been proposed to articulate constitutional rights for social media. This phenomenon has been described in terms of the emergence of a movement of 'digital constitutionalism'.

Keywords Normative dilemma • Private norms • Authoritarianism • Imperialism • Anomie • Constitutionalisation • Bills of rights • Digital constitutionalism

E. Celeste et al., *The Content Governance Dilemma*, Information Technology and Global Governance,
https://doi.org/10.1007/978-3-031-32924-1_2

2.1   From Content Moderation
to Content Governance

In 2009 Meta CEO Mark Zuckerberg said:

> More than 175 million people use Facebook. If it were a country, it would be the sixth most populated country in the world. (Zittrain 2009)

Less than two decades later, the similitude between social media platforms and nations no longer works. Today Facebook has 2.9 billion monthly active users, which means more than twice the number of inhabitants of the most populous country in the world, China (Statista 2022; World Population Review 2022). However, this analogy is still helpful to understand recent governance and regulatory trends that have characterised the sector in the past few years. One of these is a phenomenon of progressive institutionalisation. As it occurred in the context of development of the nation state, social media platforms too have gradually introduced internal norms, procedures and mechanisms to address increasingly complex issues involving a significant number of users (Sanders 2006).

In 2015, Carr and Hayes proposed a future-proof definition of social media platforms as "Internet-based, disentrained, and persistent channels of masspersonal communication facilitating perceptions of interactions among users, deriving value primarily from user-generated content" (Carr and Hayes 2015, 49). What would distinguish social media such as Facebook, Instagram, Tinder and YouTube from other online services such as emails, news websites, Zoom or Wikipedia are six main factors (Carr and Hayes 2015). Firstly, social media are not necessarily Web-based; users can access them by simply relying on an Internet connection without having to access a World Wide Web browser, as is the case while using apps like Tinder. Secondly, social media are characterised by 'disentrainment' (Carr and Hayes 2015, 50): communications over them occur in an asynchronous way, without the need of putting in place 'entrainment' mechanisms that would push and facilitate synchronous exchange. This is because social media are persistent channels; they do not disappear when users are not online but rather offer the possibility to connect at any time and resume the flow of the conversation. Thirdly, the boundaries between interpersonal and mass communication are blurred on social media: hence, Carr and Hayes' reference to 'masspersonal communication' (Carr and Hayes 2015, 52; O'Sullivan and Carr 2018). Users employ social media as an instrument of both interpersonal and mass communication without a neat demarcation between these two dimensions. Moreover, despite their asynchronous nature, users constantly have a 'perception of' interaction (Carr and Hayes 2015, 51). In other words, users might not be interacting with each other directly, in an interpersonal way, but the technical environment created by social media might provide a perception of interaction, as is the case when one is able to identify users located in a specific geographical area on Tinder. However, in ultimate analysis the added value of using social media would not lie in the content generated by platforms but by users themselves.

Indeed, social media platforms first appeared in the mid-1990s with the commercialisation of the Internet but proliferated only in the early 2000s (van Dijck 2013). For the first time in the history of the Internet, social media allowed users to be at the same time producers and consumers of the content published online. The blurring of the traditional distinction between Internet content creators and users determined the emergence of 'prosumers' on social media platforms and marked the beginning of a second phase of the Web, the so-called Web 2.0 (Fuchs 2011). However, if at the beginning the users themselves were able to moderate the content published on social media platforms, this reality soon became a utopia due to the sharp increase of users and content published (Gorwa et al. 2020). Companies managing social media platforms had to step in and introduce general rules and mechanisms to screen, assess and possibly remove the content published online in order to hinder forms of harm and abuse (Flew et al. 2019; Grimmelmann 2015). Online content moderation transitioned from community- to company-led, and unavoidably became part of these organisations' commercial activities (Gorwa et al. 2020). Moderators were no longer volunteers drawn from the cohort of users. Companies had to hire an increasing number of staff members to deal with the titanic volume of content generated online by users every day. The 'wisdom' of the community that until then had informed a bespoken interpretation of social media moderation rules was replaced with general standard guidelines to be implemented in an invariable and uniform way by external professionals. Metaphorically speaking, this radical transformation represented the transition from craftsmanship to industry in online content moderation. It is at this point of the history of online content moderation on social media platforms that one can observe the emergence of proper content *governance* systems (Gorwa 2019a, 2019b).

2.2   Micro and Macro Governance Tensions

The growth of social media platforms required the adoption of standardised rules, the institutionalisation of content review mechanisms and the professionalisation of the actors involved in moderating content online. The internal norms and structures which were consequently established progressively defined a first layer of content governance that in this book we will call 'micro' governance, as opposed to a 'macro' governance dimension which is represented by the mechanisms developed in conjunction with external actors, such as governments and advocacy groups, at a more general level (Gorwa 2019b). In the same way, it has been commonly distinguished between governance (and regulation) *by* platforms and *of* platforms (Gillespie 2018b).

Indeed, the increased centrality of online platforms in the daily life of individuals, the role played by social media in terms of allowing individuals to exercise essential freedoms and the associated level of risk of fundamental rights violations on social media transformed online platforms from mere actors of regulation to subjects of regulation. States progressively changed their approach to social media platforms. If originally these organisations were treated as mere intermediaries of information online, thus enjoying a limitation of liability for the content published by their users, over the past few years there has been an increasing tendency to recognise the role that these entities can play in limiting fundamental rights violations online (Frosio 2020, 2022). National and supranational regulators are therefore progressively shifting towards a model of co-regulation where social media platforms are entrusted the responsibility to monitor the content published by their users and to promptly intervene in order to prevent fundamental rights infringements deriving from a broad array of behaviours sanctioned by the law, from hate speech to incitement to violence (Iglesias Keller 2022).

Micro and macro content governance systems are not mutually exclusive, yet their degree of complementarity has still to be improved. The main tensions between these two governance layers are generated by two factors: the blurred boundaries between the private and public dimensions of the social media ecosystem, and the unavoidable fragmentation of the state regulatory response at global level.

The private-public distinction should theoretically inform the rationale behind the delimitation of the reciprocal actions of the micro and macro governance systems. For instance, it should demark where social media, from being mere private spaces of interaction that can be autonomously governed, assume a public relevance, and state regulation might be thus needed to enforce fundamental rights and prevent potential violations (Gillespie 2018a; Jørgensen and Zuleta 2020). However, these private online spaces have today acquired a public, not to say 'constitutional' relevance (Celeste 2021a; Celeste et al. 2022a). Individuals spend an increasing amount of their life on social media. It is no longer possible to neatly distinguish between physical and virtual life of a person as the latter is complementary to the first one and vice versa. Our physical life would not be the same without our virtual interactions so much that the digital world can be regarded as an integral component of the context where we live (Dowek 2017; Karppi 2018). Today one could no longer think of exercising some of our core fundamental rights without resorting to social media. Communicating, acquiring information, expressing our political or religious faith, protesting and exercising our businesses are only some examples of fundamental liberties that we would not be able to enjoy at the same standard if deprived of the use of social media platforms. It is certainly possible to exercise these rights in an 'analogue' way but digital technology, and in particular social media, has definitely increased the standard to which we are accustomed to exercise these rights.

In 2017 the US Supreme Court in the seminal judgement *Packingham v. North Carolina* recognised social media as "the most powerful mechanisms available to a private citizen to make his or her voice heard" (Packingham v. North Carolina 2017, 8; Celeste 2018, 2021a). Yet, at the same time, these modern public squares are owned and managed by private organisations, which are legally entitled to pursue their business interests and autonomously regulate their platforms. Contemporary German case law speaks of a *virtuelles Hausrecht*, literally the right of the digital householder, recognising the ability of platforms of banning users contravening their internal rules from the virtual domains (Celeste 2021a). Along the same lines, a common similitude employed with regard to social media links these organisations with feudal systems (Schneier 2013; Jensen 2020; Lehdonvirta 2022). These platforms create and manage autonomous virtual spaces with the power of arbitrarily defining their internal rules, as medieval dignitaries used to do in their fiefdoms.

Yet—and here the historical metaphor holds true again—online platforms do not represent virtual entities suspended in a legal vacuum, but these companies operate in physical jurisdictions. Their intangible territories host the legal and illegal actions of flesh and blood users that live in the real world. Feudalism was characterised by a multi-layered system of governance: the king of England was a vassal of the king of France; the emperor of the Holy Roman Empire revendicated power on his constituent kingdoms; the pope claimed authority on all religious affairs regardless of the existence of other personal or geographical connections to a territory that was not subject to his temporal power (Maiolo 2007). The legal maxim *rex imperator in regno suo*, the king is emperor in his kingdom, meaning that he can exercise full sovereignty and power (*plenitudo potestatis*) within the boundaries of his territories, was only introduced at the end of the Middle Ages to support the ambitions of emerging nation states, such as the Kingdom of France (Jostkleigrewe 2018). Similarly, the micro governance by social media platforms is subject to the constraints developed by the macro governance mechanisms introduced by external stakeholders, among which the one being the most impactful is state regulation.

However, while micro governance by social media platforms is unitary in nature, in the sense that each governance system at this level represents a coherent and self-sufficient entity, macro governance mechanisms are plural. The monadic unity of platforms' internal rules has to cope with the multiplicity of legal obligations originating from the various national and supranational systems in which the social media is accessible. This asymmetry generates the second element of friction between micro and macro governance. Not only do micro governance systems clash with the public objectives and values of the virtual space that social media represent for the society, but what should theoretically guide them in recomposing this tension, that is, the action of the state under the form of legal regulation, is not unitary, as many are the states and jurisdictions simultaneously affected by a single virtual social media space.

2.3   A Normative Dilemma

Micro and macro governance tensions generate a complex normative dilemma for social media companies (Celeste et al. 2022b).[1] The central question is: Which rules should govern content online? Private norms, which would ensure coherence at platform level but are arbitrarily determined by the companies themselves, or democratically voted laws? And if

[1] Klonick spoke of the 'impossible problem' of adopting a global norm regulating online speech (2019, 2427).

more national laws or international standards are simultaneously applicable to one single social media virtual space, which extends across various countries around the globe, which law to choose? How to avoid the risk of having one national or international approach imperialistically imposed on the others without resorting to third and more neutral private norms of social media companies?

Online content governance is currently facing a problem which is not novel in its essence. Determining which principles govern global spaces is an issue that characterised all phenomena related to globalisation and has affected the Internet since its origin. In his seminal book 'Code 2.0', Lessig schematised this dilemma as being the choice between a 'no law', 'one law' and 'many laws' worlds (Lessig 2006). In the social media environment, the decision of private platforms to adopt their own internal rules has been accused of arbitrariness and lack of accountability, being even associated with a 'no law' scenario (Suzor 2019). Yet, this choice is not only justified by the legal qualification of social media companies, which are private companies and are therefore legally entitled to define the rules that define their private spaces, but also by the legal pluralism that characterises national and international law. By adopting their own internal rules, social media companies are bypassing a twofold issue: firstly, the problem of reconciling multiple overlapping sets of legislation that might be simultaneously applicable to one single social media platform and, secondly, the problem of choosing the law of one country or one group of countries among many. This dilemma exposes a tension between the risk of normative authoritarianism, imperialism and anomie.

2.3.1 Authoritarianism

Pozen defined Facebook's way of establishing its own content moderation rules as a form of 'authoritarian constitutionalism' (Pozen 2018). As recognised by Celeste, online platforms' terms of service represent private constitutions as they regulate the exercise of users' rights in these virtual spaces (2019b; Suzor 2016). Social media companies have the power to unilaterally establish and amend their terms of service with no need to ensure transparency or democratic legitimacy, as in an 'absolutist' regime (Pozen 2018). According to Pozen, the internal rules of private platforms would still represent an expression of constitutionalism, as they generally try to promote values and principles, such as freedom of expression, which derive from the contemporary constitutionalist doctrine (Tushnet 2019).

Yet, the formulation, articulation and implementation of constitutional principles are in the hands of a single decision-maker, the social media company in question. De Gregorio posits that from a constitutional perspective there is no separation of powers in the field of online content governance: the prerogatives to make rules, interpret and enforce them are in the hands of the same actor (De Gregorio 2019). The same author highlights the 'paradoxical' aspect of the internal rules established by social media platforms: they are formally inspired and indeed resort to the terminology and rhetoric of constitutional values, but are de facto guided by the private interests of these commercial entities (De Gregorio 2020).

From a legal perspective, platforms' terms of service are contracts between private parties: the social media company, on the one hand, and the user, on the other hand. However, such a description of social media terms of service is extremely formal and reductive. Firstly, online platforms' internal rules are non-negotiable; they constitute 'boilerplate' contracts, which users have no choice than accepting if they want to access the social media virtual space (De Gregorio 2019; Venturini et al. 2016). Secondly, given the role de facto played by these contracts, terms of service are more similar to law, as they are norms of general application that affect millions of individuals worldwide. The scholarship has indeed talked of *lex Facebook* (Bygrave 2015) or *lex digitalis* (Karavas and Teubner 2005; Teubner 2017; Celeste 2022a) to denote the law imposed by social media platforms. Social media platforms not only rule in a 'softer' (York and Zuckerman 2019)[2] or, we would rather argue, in a more concealed way, through their technology, the algorithms that determine the content users will be 'fed' with, what Lessig called the 'code' (Lessig 2006) and Reidenberg the 'lex informatica' (Reidenberg 1998). The companies also become '*legis*-lators', literally 'promoters of the law', and this time 'law' in its traditional sense, as a set of norms expressed in words. Teubner uses the concept of *lex electronica*, which would represent an application of the notion of *lex mercatoria* to the digital field (Teubner 2004). However, the *lex electronica* is not only comparable with the ordinary law of the various sub-sectors that compose the contemporary digital society, but would represent their constitution. According to Teubner, a series of 'civil constitutions' emerge beyond the state dimension, defining the

---

[2] York and Zuckerman draw a distinction between soft and hard control, the first based on the ability of using algorithms that can determine what the users can see on the platforms and the second consisting in the platforms' content self-regulation via the terms of service.

constitutional affordances of the actors of various specific societal sub-sectors (Teubner 2012). Building on Teubner, Celeste defines social media's terms of service as constitutional instruments emerging outside the state-centric dimension, not only in light of their ability to affects users' fundamental rights on online platforms but also due to their potential role as self-restraining norms for social media companies themselves, despite their connatural limited use in this sense, as observed by Suzor (Celeste 2019b, 2022a; Suzor 2018). This form of constitutionalisation occurs in a space at least originally left outside the regulatory spectrum of nation states, relying on the capacity of online platforms of regulating themselves, establishing the rules that govern speech in their virtual spaces (Belli and Venturini 2016). Social media content moderation policies are at the same time "the most important editorial guide sheet the world has ever created", as Miller put it (qtd in Solon 2017); a contract whose force is even stronger than the law (Belli and Venturini 2016); private statutes that apply transnationally to millions of users (Langvardt 2018); and constitutional instruments regulating the exercise of fundamental rights online (Celeste 2019b; Teubner and Fischer-Lescano 2004).

2.3.2   Imperialism

If the adoption of social media's own values and principles has been accused of representing a 'no law' scenario or a form of non-democratic authoritarianism, the solution of resorting to the law of one specific country does not appear a better one either. At first sight, legally speaking, this might seem the most effective and easy to implement mechanism for a private company incorporated in one specific country to comply with the law of that state and to promote the adoption of democratically legitimated rules. However, this 'one law' scenario conceals the risk of incurring in a form of normative imperialism that does not fit the transnational and plural virtual space of social media. Indeed, most of the major social media companies are incorporated in the United States. Adopting US fundamental rights standards for content moderation would imply a forced harmonisation of the plurality of approaches to the issue of balancing freedom of speech against competing rights and interests that characterise jurisdictions around the globe. The US legal tradition, in particular, is significantly protective of the individuals' freedom of expression, enshrined in the First Amendment to the US Constitution (Pollicino 2019; Krotoszynski 2006). This would imply a limitation of the possibilities to moderate content published on social media, given the prevalence of the individual's freedom of speech over other competing interests.

Users of jurisdictions where freedom of expression is balanced in a more equal way with other rights and values would find themselves to act in a social media ecosystem regulated by rules that they would not be accustomed to, and that might be far from their legal tradition, conception of justice and culture more generally (Sangsuvan 2014). Moreover, given the fact that most social media companies are incorporated in the United States, this would also mean that a Western, US-centric prominence, which is already a matter of fact in many fields, would be perpetuated in the social media environment (Baym 2015). In a time where digital sovereignty claims are progressively emerging to contrast the de facto economic and legal imperialism of US and Chinese corporations, which share the monopoly of the tech sector, the adoption of a US-dominating 'one law' solution appears even more problematic (Celeste 2021b). Indeed, particularly in the European Union, a new conception akin to digital autarchy aiming to protect the European fundamental rights model and to emancipate member states' shared market from the predominance of the US and Chinese tech products and services is emerging and thriving, at times pushed by a sovereigntist rhetoric (Floridi 2020).

2.3.3   Anomie

In light of the risks of a 'lawless' social media environment regulated by rules arbitrarily established by online platforms (Suzor 2019), or one imperialistically dominated by the legal conception of a single country, the solution vocally invoked in the past few years has been to ensure that the content moderation rules included in the terms of service be in line with international human rights law (ARTICLE 19 2018).

At first sight, this option seems to follow the traditional legal approach of resorting to international law when transnational challenges demand to address global issues. However, this position is de facto weakened by the legal reality that does not know the existence of a single human rights law standard, but conversely a plurality of legal models, interpreted differently by courts and professionals around the globe (Mégret 2013).[3] Particularly in the field of freedom of expression, international human rights law

---

[3] More in general on the issue of fragmentation of international law, see vol. 25, issue 4 of the *Michigan Journal of International Law* (2004) 845 ff.

exposes divergent approaches, often as a heritage of national constitutional traditions. Therefore, this would mean that behind the claim that it would suffice to bring private content moderation standards in line with international human rights law the issue of choosing one legal approach among many persists, similarly to what happens if social media companies decided to apply the law of one country. The problem of determining which legal standards to apply does not move away, together with its connatural issues of potential normative imperialism and distance from the cultural pluralism that characterises the social media environment.

Secondly, one particular issue that characterises international human rights law, making it less suitable to govern online content moderation, is that its norms traditionally address states and not private actors. Legally speaking, international law obligations only bind states. There are international instruments advocating for an increased responsibility of private actors in ensuring that their activities do not infringe fundamental rights, also through preliminary risk and impact assessments; yet these documents only have value of soft law.[4] Moreover, this discrepancy between the traditional addressees of international law and the dominant actors of online content moderation generates complexities in terms of interpretation of these norms. International human rights standards are not directly applicable to online content moderation cases as they require a work of legal interpretation and recontextualisation.

An issue that is made even more problematic in light of the lack of granularity of international human rights norms. At the international level, indeed, fundamental rights and liberties are framed as general principles. There are neither provisions tailored to the social media environment nor specific mechanisms created to operate a balancing of competing rights and interests in the context of online content moderation. This circumstance exposes an issue of potential normative anomie, a sense of disorientation that emerges in the phase of implementation of norms to concrete content moderation cases (Celeste 2022a). International human rights standards, by defining general orienting principles, require a substantial degree of interpretation and, unavoidably, a sufficient legal knowledge.

---

[4] For example, the UN Guiding Principles on Business and Human Rights (UNGPs). (See also Kaye 2018). See also David Kaye, "Report of the Special Rapporteur on the Promotion and Protection of the Right to Freedom of Opinion and Expression—A/HRC/38/35". https://documents-dds-ny.un.org/doc/UNDOC/GEN/G18/096/72/PDF/G1809672.pdf?OpenElement. For a more detailed analysis of this point, see *infra* Chap. 3.

These norms, as they are, could not offer explicit guidance of behaviour to the actors involved in the social media environment and could not be directly applicable without a preliminary work of interpretation and recontextualisation (Belli and Venturini 2016).[5] Arguing that international human rights law would be the panacea of the online content governance dilemma is a false myth.

2.4   THE POTENTIAL OF DIGITAL CONSTITUTIONALISM

A straightforward solution to the question of which standards should govern online content moderation cannot be represented by a mere legal transplant. Adopting the law of one country or referring to international human rights standards are options that conceal a series of significant problems, in particular due to the fact these legal frameworks were not intended to govern a transnational and plural environment dominated by private actors like the one of online platforms. A twofold work of translation and adaptation is needed in order to ensure that social media standards comply with fundamental rights. On the one hand, international and national norms enshrining fundamental rights principles have to be recontextualised in light of the specificities of the social media environment, and, on the other hand, platform standards must be reshaped in order to progressively incorporate these values. A process of 'constitutionalisation' of the social media environment seems to be needed (Celeste et al. 2022a; Celeste 2022a, Chap. 5). Instilling the core principles of contemporary constitutionalism in the architecture of social media would mean to preserve the legal effectiveness of platform standards while enhancing their capability to promote the respect of fundamental rights in the multinational and plural environment they govern.

Interestingly, an input to this process of constitutionalisation is increasingly originating from civil society actors. Over the past few years, a significant number of 'declarations' or 'bills of rights' have been proposed to articulate constitutional rights and principles in a way that would reflect and address the challenges of the digital age (Redeker et al. 2018; Yilma 2021; Celeste 2022a). This phenomenon has been described in terms of emergence of a movement of 'digital constitutionalism' (Redeker et al. 2018; Padovani and Santaniello 2018; Suzor 2018; Celeste 2019a;

---

[5] Cf. also in this text the suggestion of the authors of adopting international technical standards, such as ISO 26000, to facilitate multistakeholder participation and accountability.

Pollicino 2021; De Gregorio 2021; Celeste 2022a). These documents do not represent legally binding texts, yet they often adopt the 'lingua franca' of constitutional law (Celeste 2022a). Singularly taken, the contribution of these documents from a constitutional law perspective is limited. However, regarded as a comprehensive movement composed by a plurality of initiatives, these civil society efforts have so far nourished a conversation on which values and principles should govern the digital ecosystem (Celeste 2022a, Chap. 8). These declarations promote an update and re-articulation of core principles of contemporary constitutionalism, rather than a complete re-writing of norms. They do not aim to subvert the DNA of contemporary constitutionalism, but rather to 'generalise and respecify' its core values in light of the mutated social reality where we live (Celeste 2022b).

In this context, the social media environment emerges as a laboratory of new ideas. Internet bills of rights often include principles that explicitly address common challenges of online content moderation and could help develop platform standards that are more in line with fundamental rights. These documents represent a voice often unheard. The closest one to the users, whose opinion is way far neglected in the context of the content governance dilemma, being them subject to private standards or public laws without having the possibility to express how they think their fundamental freedoms should be articulated and balanced on online platforms.

Internet bills of rights do not claim to become cosmopolitan constitutions for the social media environment but provide an impulse to the conversation on how to instil constitutional values within online platforms standards. Despite the evocative image that the concept of 'constitutionalisation' brings to mind, there are no founding fathers—or mothers—sitting in the same room for days that aim to define a single constitution for social media. The process of constitutionalisation of this environment reflects the complex, global and plural scenario in which online platforms operate. Online content governance rules are being fertilised by a multi-stakeholder constitutional input. An aerial view on this phenomenon witnesses multiple, simultaneous processes of 'parallel' or 'collateral' constitutionalisation that are currently ongoing (Celeste et al. 2022b; Celeste 2019a, 2022a, 2022b). Civil society's Internet bills of rights may be regarded as one of the inputs that contribute to shaping this plural phenomenon. The online content governance dilemma might not be solved by choosing to stick to private rules *or* refer to national or international law: the ultimate solution might be a *combination* of these options.

The conversation on digital constitutionalism is polyphonic, nourished by a plurality of voices, also emerging from below. Internet bills of rights work as a linking element that help connect, complement and stimulate these various normative dimensions to find answers to the challenges of online platforms (Celeste 2022a, Chap. 8).

REFERENCES

ARTICLE 19. 2018. *Facebook Community Standards: June 2018—A Legal Analysis.* https://www.article19.org/wp-content/uploads/2018/07/Facebook-Community-Standards-August-2018-1-1.pdf.

Baym, Nancy K. 2015. Social Media and the Struggle for Society. *Social Media PLUS_SPI Society* 1 (1): 205630511558047.

Belli, Luca, and Jamila Venturini. 2016. Private Ordering and the Rise of Terms of Service as Cyber-Regulation. *Internet Policy Review* 5 (4).

Bygrave, Lee A. 2015. Lex Facebook. In *Internet Governance by Contract.* Oxford: Oxford University Press.

Carr, Caleb T., and Rebecca A. Hayes. 2015. Social Media: Defining, Developing, and Divining. *Atlantic Journal of Communication* 23 (1): 46–65.

Celeste, Edoardo. 2018. Packingham v North Carolina: A Constitutional Right to Social Media? *Cork Online Law Review* 17: 116–119.

———. 2019a. Digital Constitutionalism: A New Systematic Theorisation. *International Review of Law, Computers & Technology* 33 (1): 76–99.

———. 2019b. Terms of Service and Bills of Rights: New Mechanisms of Constitutionalisation in the Social Media Environment? *International Review of Law, Computers & Technology* 33 (2): 122–138.

———. 2021a. Digital Punishment: Social Media Exclusion and the Constitutionalising Role of National Courts. *International Review of Law, Computers & Technology* 35: 162–184.

———. 2021b. Digital Sovereignty in the EU: Challenges and Future Perspectives. In *Data Protection Beyond Borders: Transatlantic Perspectives on Extraterritoriality and Sovereignty*, ed. Federico Fabbrini, Edoardo Celeste, and John Quinn, 211–228. Oxford: Hart.

———. 2022a. *Digital Constitutionalism: The Role of Internet Bills of Rights.* Abingdon, Oxon; New York, NY: Routledge.

———. 2022b. The Constitutionalisation of the Digital Ecosystem: Lessons from International Law. In *Digital Transformations in Public International Law*, ed. Angelo Golia, Matthias C. Kettemann, and Raffaela Kunz. Baden-Baden: Nomos.

Celeste, Edoardo, Amélie Heldt, and Clara Iglesias Keller, eds. 2022a. *Constitutionalising Social Media.* Oxford: Hart.

Celeste, Edoardo, Kinfe Micheal Yilma, Nicola Palladino, and Dennis Redeker. 2022b. Digital Constitutionalism: In Search of a Content Governance Standard. In *Constitutionalising Social Media*, ed. Edoardo Celeste, Amélie Heldt, and Clara Iglesias Keller. Oxford: Hart.

De Gregorio, Giovanni. 2019. From Constitutional Freedoms to the Power of the Platforms: Protecting Fundamental Rights Online in the Algorithmic Society. *European Journal of Legal Studies* 11 (2): 65–103.

———. 2020. Democratising Online Content Moderation: A Constitutional Framework. *Computer Law and Security Review* 36: 105374.

———. 2021. The Rise of Digital Constitutionalism in the European Union. *International Journal of Constitutional Law* 19 (1): 41–70.

Dowek, Gilles. 2017. *Vivre, aimer, voter en ligne et autres chroniques numériques.* Paris: Le Pommier.

Flew, Terry, Fiona Martin, and Nicolas Suzor. 2019. Internet Regulation as Media Policy: Rethinking the Question of Digital Communication Platform Governance. *Journal of Digital Media & Policy* 10 (1): 33–50.

Floridi, Luciano. 2020. The Fight for Digital Sovereignty: What It Is, and Why It Matters, Especially for the EU. *Philosophy & Technology* 33 (3): 369–378.

Frosio, Giancarlo, ed. 2020. *Oxford Handbook of Online Intermediary Liability.* Oxford University Press.

———. 2022. Regulatory Shift in State Intervention: From Intermediary Liability to Responsibility. In *Constitutionalising Social Media*, ed. Edoardo Celeste, Clara Iglesias Keller, and Amélie Heldt, 151–176. Hart.

Fuchs, Christian. 2011. Web 2.0, Prosumption, and Surveillance. *Surveillance and Society* 8 (3): 288–309.

Gillespie, Tarleton. 2018a. *Custodians of the Internet: Platforms, Content Moderation, and the Hidden Decisions that Shape Social Media.* New Haven: Yale University Press.

———. 2018b. Regulation of and by Platforms. In *The SAGE Handbook of Social Media*, ed. Jean Burgess, Alice Marwick, and Thomas Poell, 254–278. Sage.

Gorwa, Robert. 2019a. The Platform Governance Triangle: Conceptualising the Informal Regulation of Online Content. *Internet Policy Review* 8 (2).

———. 2019b. What Is Platform Governance? *Information, Communication & Society* 22 (6): 854–871.

Gorwa, Robert, Reuben Binns, and Christian Katzenbach. 2020. Algorithmic Content Moderation: Technical and Political Challenges in the Automation of Platform Governance. *Big Data & Society* 7 (1).

Grimmelmann, James. 2015. The Virtues of Moderation. *Yale Journal of Law and Technology* 17: 42–109.

Iglesias Keller, Clara. 2022. The Perks of Co-Regulation: An Institutional Arrangement for Social Media Regulation? In *Constitutionalising Social Media*, ed. Edoardo Celeste, Clara Iglesias Keller, and Amélie Heldt, 217–233. Hart.

Jensen, Jakob Linaa. 2020. *The Medieval Internet: Power, Politics and Participation in the Digital Age.* 1st ed. Bingley, UK: Emerald Publishing Limited.

Jørgensen, Rikke Frank, and Lumi Zuleta. 2020. Private Governance of Freedom of Expression on Social Media Platforms: EU Content Regulation through the Lens of Human Rights Standards. *Nordicom Review* 41 (1): 51–67.

Jostkleigrewe, Georg. 2018. 'Rex Imperator in Regno Suo': An Ideology of Frenchness? Late Medieval France, Its Political Elite and Juridical Discourse. In *Imagined Communities: Constructing Collective Identities in Medieval Europe*, ed. Andrzej Pleszczyński, Joanna Aleksandra Sobiesiak, Michał Tomaszek, and Przemysław Tyszka, 46–83. Brill.

Karavas, Vagias, and Gunther Teubner. 2005. www.CompanyNameSucks.Com: The Horizontal Effect of Fundamental Rights on 'Private Parties' Within Autonomous Internet Law. *Constellations* 12 (2): 262–282.

Karppi, Tero. 2018. *Disconnect.* Minneapolis; London: University of Minnesota Press.

Kaye, David. 2018. Report of the Special Rapporteur on the Promotion and Protection of the Right to Freedom of Opinion and Expression—A/HRC/38/35. https://documents-dds-ny.un.org/doc/UNDOC/GEN/G18/096/72/PDF/G1809672.pdf?OpenElement.

Klonick, Kate. 2019. The Facebook Oversight Board: Creating an Independent Institution to Adjudicate Online Free Expression. *The Yale Law Journal* 129 (8): 2418–2499.

Krotoszynski, Ronald J. 2006. *The First Amendment in Cross-Cultural Perspective: A Comparative Legal Analysis of the Freedom of Speech.* New York University Press.

Langvardt, Kyle. 2018. Regulating Online Content Moderation. *Georgetown Law Journal* 106: 1353–1388.

Lehdonvirta, Vili. 2022. *Cloud Empires: How Digital Platforms Are Overtaking the State and How We Can Regain Control.* Cambridge, Massachusetts: The MIT Press.

Lessig, Lawrence. 2006. *Code: And Other Laws of Cyberspace, Version 2.0.* New York: Basic Books.

Maiolo, Francesco. 2007. *Medieval Sovereignty: Marsilius of Padua and Bartolus of Saxoferrato.* Delft: Eburon.

Mégret, Frédéric. 2013. International Human Rights and Global Legal Pluralism: A Research Agenda. In *Dialogues on Human Rights and Legal Pluralism*, Ius Gentium: Comparative Perspectives on Law and Justice, ed. René Provost and Colleen Sheppard, 69–95. Dordrecht: Springer Netherlands.

O'Sullivan, Patrick B., and Caleb T. Carr. 2018. Masspersonal Communication: A Model Bridging the Mass-Interpersonal Divide. *New Media & Society* 20 (3): 1161–1180.

Packingham v. North Carolina. 2017. (US Supreme Court).

Padovani, Claudia, and Mauro Santaniello. 2018. Digital Constitutionalism: Fundamental Rights and Power Limitation in the Internet Eco-System. *International Communication Gazette* 80 (4): 295–301.

Pollicino, Oreste. 2019. Judicial Protection of Fundamental Rights in the Transition from the World of Atoms to the Word of Bits: The Case of Freedom of Speech. *European Law Journal* 25 (2): 155–168.

———. 2021. *Judicial Protection of Fundamental Rights on the Internet: A Road Towards Digital Constitutionalism?* Oxford: Hart.

Pozen, David. 2018. Authoritarian Constitutionalism in Facebookland. http://knightcolumbia.org/content/authoritarian-constitutionalism-facebookland. Accessed October 13, 2022.

Redeker, Dennis, Lex Gill, and Urs Gasser. 2018. Towards Digital Constitutionalism? Mapping Attempts to Craft an Internet Bill of Rights. *International Communication Gazette* 80 (4): 302–319.

Reidenberg, Joel. 1998. Lex Informatica: The Formulation of Information Policy Rules through Technology. *Texas Law Review* 76 (3): 553–593.

Sanders, Elizabeth. 2006. Historical Institutionalism. In *The Oxford Handbook of Political Institutions*, ed. R.A.W. Rhodes, Sarah A. Binder, and Berth A. Rockman, 39–55. Oxford University Press.

Sangsuvan, Kitsuron. 2014. Balancing Freedom of Speech on the Internet Under International Law. *North Carolina Journal of International Law* 39: 701–755.

Schneier, Bruce. 2013. Power in the Age of the Feudal Internet. Ed. Wolfgang Kleinwächter. *MIND* (#6-'Internet and Security'): 16–21.

Solon, Olivia. 2017. To Censor or Sanction Extreme Content? Either Way, Facebook Can't Win | Facebook. *The Guardian*. https://www.theguardian.com/news/2017/may/22/facebook-moderator-guidelines-extreme-content-analysis. Accessed October 14, 2022.

Facebook MAU Worldwide 2022. *Statista*. https://www.statista.com/statistics/264810/number-of-monthly-active-facebook-users-worldwide/. Accessed September 23, 2022.

Suzor, Nicolas. 2016. The Responsibilities of Platforms: A New Constitutionalism to Promote the Legitimacy of Decentralized Governance. Berlin. http://eprints.qut.edu.au/101953/. Accessed October 13, 2022.

———. 2018. Digital Constitutionalism: Using the Rule of Law to Evaluate the Legitimacy of Governance by Platforms. *Social Media PLUS_SPI Society* 4 (3): 1–11.

———. 2019. *Lawless. The Secret Rules That Govern Our Digital Lives.* Cambridge: Cambridge University Press.

Teubner, Gunther. 2004. Societal Constitutionalism; Alternatives to State-Centred Constitutional Theory? In *Transnational Governance and Constitutionalism. International Studies in the Theory of Private Law*, ed. Christian Joerges, Inger-Johanne Sand, and Gunther Teubner, 3–28. Oxford; Portland: Hart.

————. 2012. *Constitutional Fragments: Societal Constitutionalism and Globalization*. Oxford: Oxford University Press.

————. 2017. Horizontal Effects of Constitutional Rights in the Internet: A Legal Case on the Digital Constitution. *The Italian Law Journal* 3 (1): 193–205.

Teubner, Gunther, and Andreas Fischer-Lescano. 2004. Regime-Collisions: The Vain Search for Legal Unity in the Fragmentation of Global Law. *Michigan Journal of International Law* 25: 999–1046.

Tushnet, Rebecca. 2019. Content Moderation in an Age of Extremes. *Journal of Law, Technology and the Internet* 10 (1): 1–19.

van Dijck, José. 2013. *The Culture of Connectivity: A Critical History of Social Media*. Oxford; New York: Oxford University Press.

Venturini, Jamila, et al. 2016. *Terms of Service and Human Rights: An Analysis of Online Platform Contracts*. Editora Revan.

Total Population by Country 2022. *World Population Review*. https://worldpopulationreview.com/countries.

Yilma, Kinfe. 2021. Bill of Rights for the 21st Century: Some Lessons from the Internet Bill of Rights Movement. *The International Journal of Human Rights* 26 (4): 701–716.

York, Jillian C., and Ethan Zuckerman. 2019. Moderating the Public Sphere. In *Human Rights in the Age of Platforms*, ed. Rikke Frank Jørgensen, 137–161. The MIT Press.

Zittrain, Jonathan. 2009. A Bill of Rights for the Facebook Nation. *The Chronicle of Higher Education*. https://www.chronicle.com/blogs/wiredcampus/jonathan-zittrain-a-bill-of-rights-for-the-facebook-nation/4635. Accessed August 30, 2018.

CHAPTER 3

The International Law of Content Governance

Abstract This chapter considers the extent to which international law offers guidance to the governance of digital platforms. Recent years have seen a growing attention in the potential of international law in offering normative guidance to address human rights concerns in content governance. The chapter considers whether the recent turn to international law for content governance standards is a worthwhile exercise. It finds that, beset by a host of design and structural constraints, international law does not offer meaningful normative guidance to the governance of and in digital platforms. The chapter then highlights how emergent standards are moving past the regulatory limits of international law.

Keywords International human rights law • International law of content governance • Business and human rights • Emergent standards • Freedom of expression • Social media • Digital platforms

3.1 Unveiling a Myth

Recent years have seen a growing attention in the potential of international law in offering normative guidance to address human rights concerns in content governance. Partly in response to the pressure from civil society groups—including through the launching of Internet bills of rights that advance progressive content governance standards, social

E. Celeste et al., *The Content Governance Dilemma*, Information Technology and Global Governance,
https://doi.org/10.1007/978-3-031-32924-1_3

media platforms are also increasingly yet vaguely turning their attention to international human rights law (Meta 2021; Twitter 2022). As shall be shown in the next chapter, a number of civil society organisations have pushed for social media companies to ground their content moderation policies in international human rights standards. Several reports of United Nations (UN) special rapporteurs and the scholarly literature have likewise argued for platform content moderation policies and practices to be based on and guided by international human rights law. But the question of whether, and the extent to which, international law offers such guidance to the complex world of platform content governance remains. This chapter seeks to address this question. It will show that the potential of international human rights law in offering much-needed normative guidance to content governance is circumscribed by two interrelated factors.

One is that international human rights law is—by design—state-centred and hence does not go a long way in attending to human rights concerns in the private sector. This means that international human rights law relegates to national law the regulation of the private sector. Problematic about this state of affairs is that it risks leading to divergent regulatory approaches globally. The other 'design constraint' is that international human rights standards are mostly couched in general principles. This makes the principles less suited to be applied in the context of platform content moderation which requires a rather granular and dynamic system of norms. The second factor concerns a set of structural constraints that further limit the regulatory potential of international human rights law. In the rare instances where soft international law standards appear to have companies in their regulatory site, they still rely by and large on voluntary compliance and hence envisage no robust accountability mechanisms. In practice, the generic international content governance standards have not adequately been unpacked by relevant adjudicative bodies such as treaty bodies to make them fit for purpose to the present realities of content moderation. On the whole, content governance jurisprudence at the international level remains to be thin.

In this chapter, the phrase 'international law of content governance' refers to a set of international standards relating to content governance provided both in international hard and soft laws. Hard legal instruments considered include the International Covenant on Civil and Political Rights (ICCPR) and the International Convention on the Elimination of All Forms of Racial Discrimination (CERD). Certain Covenant rights, particularly the right to freedom of expression, which is at the centre of content governance, are replicated nigh verbatim in other post-ICCPR specialised human rights treaties (Convention on the Rights of the Child 1989: art 13; Convention on the Rights of Persons with Disabilities 2007: art 21). As a result, references in this chapter to ICCPR provisions would, *mutatis mutandis,* apply to corresponding provisions in those treaties.

Whereas soft legal instruments include a broad range of instruments, including the Universal Declaration of Human Rights (UDHR), the United Nations Guiding Principles on Business and Human Rights (UNGPs, alternatively referred to in this chapter as Ruggie Principles),[1] relevant Resolutions of the Human Rights Council and Joint Declarations of UN and intergovernmental mandates on freedom of expression. This means that the chapter excludes regional instruments from the purview of the 'international law' analysis, mainly because such instruments are transnational/regional in scope while issues of content governance are inherently universal. International law is the most pertinent framework of reference for addressing such a universal issue.

The rest of the chapter develops in four sections. We first map the key normative sources of international law of content governance (Sect. 3.2) which consists of general and specific standards applicable to the governance of and in digital platforms. Emergent standards developed through intergovernmental mandates on freedom of expression are then considered to highlight recent norm progressions in international law (Sect. 3.3). In Sect. 3.4, we explore the ways in which a host of design and structural constraints undercut the regulatory potential of international law of content governance. We close the chapter in Sect. 3.5 where the growing gap-filling role of civil society initiatives is flagged, a subject explored in full in Chap. 4.

3.2 Normative Sources

Content governance standards in international law draw from multiple normative sources and take various legal forms. While some are embodied in hard law and hence carry binding legal obligations, others are envisaged in soft law instruments with no enforceable obligations. Whereas certain standards are general in formulation and scope. One,

---

[1] Note that the Guiding Principles or the Ruggie Principles are referred to in the singular in this chapter deliberately so as to capture the fact that it is a single instrument.

for instance, finds general norms that define the scope of human rights obligations against state parties to the relevant treaty. This, in turn, would include the obligations of states in regulating the content moderation practices of digital platforms. But this category also encompasses less explored binding norms that would potentially apply to digital platforms directly. Other content governance standards are specific in the sense that they have the potential to apply to particular cases of content governance. Such norms include international standards that deal with rights and principles engaged directly by content governance. A set of human rights guaranteed in widely accepted human rights treaties and soft legal instruments such as the UNGPs fall within this category. What follows maps this set of content governance norms in international law.

3.2.1   Generic Standards: Platforms as Duty-Bearers?

The ICCPR, a human rights treaty widely ratified by states—173 state parties at the time of writing[2]—provides the general framework for any consideration of content governance in international law. One way it does so is by defining the scope of state obligations vis-à-vis Covenant rights. States generally owe two types of obligations under the Covenant: negative and positive obligations (International Covenant on Civil and Political Rights 1966, art 2(1)). States' negative obligation imposes a duty to 'respect' the enjoyment of rights. As such, it requires States and their agents to refrain from any conduct that would impair or violate the exercise or enjoyment of rights guaranteed in the Covenant. States' positive obligation, on the other hand, imposes a duty to 'protect' the enjoyment of rights. This obligation thus concerns state regulation of third parties, including private actors, to ensure respect for Covenant rights. Article 2 of the Covenant stipulates states' positive human rights obligations as follows:

> [...] Each State Party to the Present Covenant undertakes to take the necessary steps, in accordance with its constitutional processes and with the provisions of the present Covenant, <u>to adopt such laws or other measures as may be necessary to give effect to the rights recognized in the present Covenant</u>. [Emphasis added]

[2] See details at https://indicators.ohchr.org/.

Each State Party to the present Covenant undertakes:

(a) To ensure that any person whose rights or freedoms as herein recognized are violated shall have effective remedy [...]'
(b) To ensure that any person claiming such a remedy shall have his rights thereto determined by competent judicial, administrative or legislative authorities, or by any other competent authority provided for by the legal system of the State, and to develop the possibilities of judicial remedy;
(c) To ensure that the competent authorities shall enforce such remedies when granted.

States' positive human rights obligation primarily concerns a duty to put in place the requisite legal and institutional framework to enable the enjoyment of rights, including means of recourse when violations occur (General Comment 31 2004: paras 6–7). Applied to content governance, this duty would mean that—where permitted by their respective domestic constitutional framework—states should enact laws that regulate the conduct of digital platforms, including content moderation policies and practices. Recent regulatory initiatives in several jurisdictions, such as Germany's Network Enforcement Act (NetzDG), are good cases in point (Germany's Network Enforcement Act 2017).

International human rights law generally does not impose obligations on non-state actors, including corporations; but there are certain exceptions to this rule. At the highest level, the Preamble of the UDHR states that 'every organ of society' shall strive to promote respect for rights guaranteed in the Declaration (Universal Declaration of Human Rights 1948, Preamble: para 8). Commentators argue that the reference to 'every organ of society' should be taken to include corporations, and their duty to 'respect' human rights (Henkin 1999, 25). Despite the fact that this proviso is in the inoperative parts of the Declaration—and the latter is not formally a binding instrument, it can be taken to foreshadow a negative obligation of non-state actors, including technology companies, to 'respect' human rights. At its core, the negative human rights obligation to respect is a duty to refrain from any act that would undermine the exercise or enjoyment of human rights.

Perhaps a more concrete version of this tendency to address non-state actors in compulsory terms is provided in the operative provisions of both the UDHR and the ICCPR. Article 30 of the UDHR and Article 5(1) of the ICCPR, respectively, read as follows:

Nothing in this Declaration may be interpreted as implying for any State, group or person any right to engage in any activity or to perform any act aimed at the destruction of any of the rights and freedoms set forth herein.

Nothing in the present Covenant may be interpreted as implying for any State, group or person any right to engage in any activity or perform any act aimed at the destruction of any of the rights and freedoms recognized herein or at their limitation to a greater extent than is provided for in the present Covenant.

A closer look at these provisions suggests that no right is bestowed upon anyone, including 'groups and persons', as well as states to impair or destruct the exercise and enjoyment of the rights guaranteed in the Declaration and the Covenant. It has been argued that the rationale for the inclusion of this provision in the UDHR—which was later replicated in the ICCPR with minor additions—was that persons who are opposed to the 'spirit of the Declaration or who are working to undermine the rights of men should not be given the protection of those rights' (Schabas 2013, 1308). Guided by the slogan "no freedom for the enemies of freedom", this provision is meant to prevent the abuse of rights (Opsahl and Dimitrijevic 1999, 648–649). Now the question is whether the prohibition of abuse of rights would equally apply to digital platforms.

Freedom of expression in international law is bestowed to individuals, and as such, companies including social media platforms are not right holders. That is not, however, the case in national legal systems such as the United States where First Amendment protection applies to social media companies (United States Supreme Court, Manhattan Community Access Corp. v. Halleck 2019).[3] But the prohibition of abuse of rights both in the Declaration and the Covenant arguably would also apply to social media companies in the sense that their policies and practices, including those relating to content moderation, must not have the effect of impairing or destructing the enjoyment of human rights. In that sense, there is a negative obligation to 'respect' human rights which requires them to refrain from measures that would affect the enjoyment of rights. This provision is general—and originally was meant to limit abuse by, as Nowak writes, "national socialists, fascists, racists and other totalitarian activities" who employ certain rights like freedom of expression to "destroy democratic

---

[3] The US Supreme Court in this case affirmed the First Amendment right of private publishers to control the content of their publications.

structures and human rights of others protected by such structures" (Nowak 2005, 112, 115).[4] But there is no reason why it would not apply to govern content policies and practices of digital platforms.

However, the tendency in such provisions to address non-state actors directly, potentially including corporations, appears to be at odds with the state-centric nature of human rights law generally. As alluded to above, international human rights law imposes obligations only on states who are parties to the underlying treaties imposing such obligations. Of course, the binding human rights norms in question are embodied in a treaty— that is, ICCPR—to which only states are, or can be, parties. And this state of affairs raises the fundamental question of how this would apply in the context of digital platforms. But the sheer fact that the provisions appear to impose binding duties, regardless of how they would be enforced, would certainly lend weight to recent arguments—considered later in this chapter—that international human rights law does, or should, apply directly to the content moderation practices of digital platforms.

Albeit in a different context, some commentators argue that Article 19 of the ICCPR imposes some duties directly on online intermediaries (Land 2019, 286, 303–304). Among such duties include respect for principles of due process and remediation. This expansive and potentially contentious reading of the provision relies on two points. One is the fact that the drafters of the Covenant had contemplated non-state actors as duty-bearers although this was not later incorporated in the final text. Non-state actors have long been considered potential duty-bearers in human rights standard-setting processes, but a range of factors frustrated any attempt to codify corporate human rights obligations in treaty law. Chief among such factors was lack of support from many developed countries and multinational corporations.

The more recent attempt to translate the Ruggie Principles into a human rights treaty is already facing similar challenges. At the sixth session of the Working Group that is currently drafting such a treaty, a number of states expressed reservations and outright opposition to the draft text as well as to the whole treaty process. The UK delegation, for instance, noted that while the draft has noble aims, it expressed scepticism that the text can gather enough political support (Open-ended Intergovernmental

---

[4] Note also Nowak's point that Article 5(1) of the ICCPR is closely related to Article 20 of the Covenant which prohibits two types of speeches: war propaganda and incitement to national, racial and religious hatred.

Working Group 2021, para 22). The United States, on the other hand, not only maintained objection to the process but also called on the Working Group to abandon it in favour of alternative approaches (Open-ended Intergovernmental Working Group 2021, para 23). That undercuts the normative value of abandoned "contemplation" among drafters of the ICCPR alluded to above by some commentators.

But a more direct reading of the duties of digital platforms in international law draws from the terms "special duties and responsibilities" in Article 19(3) of the ICCPR. It is argued that these terms imply duties of intermediaries, including digital platforms (Land 2019, 303–305). Yet the use of the terms in the provision is expressly in the context of the exercise of the right to freedom of expression and attendant grounds of restriction. It reads that "the exercise of the rights provided for in paragraph 2 of this article carries with it special duties and responsibilities". That simply means the envisioned "special duties and responsibilities" are owed by individuals to whom the right to freedom of expression and opinion are bestowed under international law. It is not apparent from the argument whether it flows from the rule present in some jurisdictions, particularly in the United States, where private publishers enjoy free speech rights. If that was the case, it would mean that the exercise of this right by intermediaries would entail "special duties and responsibilities". However, simply because non-state actors are not entitled to freedom of expression, there can be no corresponding duties on intermediaries in international law.

Overall, the upshot is that while certain standards appear to envision corporations as duty-bearers, the state-centred nature of international human rights law makes them less suited to content governance. Add to that their exceedingly generic formulation which has yet to be unpacked in practice. What follows considers whether more specific standards in international law may offer better normative guidance.

3.2.2   *Specific Standards: Applicable Human Rights Treaties*

Specific content governance standards in international law are envisaged in a series of human rights treaties and international soft law. Human rights treaties guarantee a broad range of rights that set out standards applicable to the governance of the conduct of or in digital platforms. The UN framework on business and human rights, also referred to as the Ruggie Principles, is the other specific standard potentially applicable to the governance of platforms. What follows considers the degree to which these standards offer effective normative guidance for content governance in digital platforms.

Content governance standards in human rights law take different forms. One finds, for instance, standards that prohibit certain types of speech: war propaganda, advocacy for racial, religious and national hatred and racist speech (International Covenant on Civil and Political Rights 1966, art 20; International Convention on the Elimination of All Forms of Racial Discrimination 1965, art 4). In outlawing certain types of expression, international human rights law essentially sets forth content governance standards that must be implemented by state parties to the relevant treaties, including in social media platforms. Social media companies are not bound by such international standards. But digital platforms are increasingly being required by domestic legislation to observe rules that essentially reflect international human rights and principles. Germany's Network Enforcement Act, for instance, has among its legislative objectives tackling hate speech, thereby upholding rights affected by such types of online speech (Germany's Network Enforcement Act 2017). In a way, community standards of digital platforms can be taken—at least theoretically—to be translation of domestic and international law standards. What this may further mean is that in many cases such platform policies will apply to users in jurisdictions where domestic legislation is non-existent or unenforced. But in the latter—and probably common—case, it would largely be a voluntary commitment on the part of digital platforms.

International human rights law not only guarantees the right to freedom of expression but also provides standards for permissible restrictions. This is the other source of specific content governance standards in human rights and principles. Restriction of freedom of expression will be permissible when three cumulative requirements are fulfilled: legality, necessity and legitimacy. Article 19(3) of the ICCPR provides the standards of restriction as follows:

> [...] It [freedom of expression] may therefore be subject to certain restrictions, but these shall only be such as are provided by law and are necessary: (a) for respect of the rights or reputation of others, (b) for the protection of national security or of public order or of public health or morals.

The three-part tests of legality, necessity and legitimacy are designed to address the restriction of rights by the states and their agents. But there have been several attempts to adapt the test or to formulate *sui generis* standards of human rights restrictions *by* corporations (Karavias 2013; Ratner 2001). More recently, several attempts to translate and adapt these standards to content governance have emerged. At the forefront of this effort has been the former UN Special Rapporteur on Freedom of Expression, David Kaye. In a series of reports, he argued how digital platforms can, and should, follow international human rights standards in the course of applying content governance standards. Doubtless, this offers some intellectual guidance on the human rights responsibilities of technology companies, including social media platforms. But Kaye's reports are particularly notable in two respects.

One is that they seek to elaborate on how the right to freedom of expression guaranteed in Article 19 of the ICCPR, as well as Article 19 of the UDHR, would apply to technology companies. Kaye argued that following international law standards in content moderation, as opposed to discretionary community standards, would allow corporations to make 'principled arguments' to protect the rights of users (Kaye 2019, 10). In particular, adapting international law would mean that requirements for permissible restriction of freedom of expression would need to be applied by social media companies. For example, the requirement of 'legality' would require platforms to adopt 'fixed rules' on content moderation that are publicly accessible and understandable to users (Kaye 2019, 43). His report on online hate speech similarly argues that companies should assess whether their hate speech rules infringe upon freedom of expression based on the requirements of legality, necessity and legitimacy (Special Rapporteur on Freedom of Opinion and Expression 2019, paras 46–52).

Reinventing the Ruggie Principles is the other way in which Kaye sought to adapt international standards to the platform governance context. We return to this point in the next section where we illustrate the extent to which the UN business and human rights framework may apply to the governance of and in digital platforms. But it is vital to note that such reports of the former Special Rapporteur often build on the submissions of various stakeholders, mainly civil society groups.[5] Indeed, the above highlighted 'procedural safeguards' alluded to by Kaye are widely advocated in civil society initiatives. The next chapter will examine civil

---

[5] His 2018 report on online content regulation, for instance, was the culmination of a year-long series of consultations, visits to major Internet companies and a wide range of state and civil society input. See details at the mandate's webpage https://www.ohchr.org/EN/Issues/FreedomOpinion/Pages/ContentRegulation.aspx.

society initiatives relating to content governance. But this phenomenon attests to the iterative cross-fertilisation of and convergence between civil society standards and international human rights standards.

Multistakeholder bodies have similarly attempted to adapt human rights standards to content governance.[6] A good case in point is the Global Network Initiative (GNI) which introduced Principles on Freedom of Expression and Privacy (GNI Principles 2008, as updated in 2017). GNI is a multistakeholder body established to serve as a platform for addressing issues relating to digital privacy and free speech through dialogue among stakeholders.[7] Although focused only on privacy and freedom of expression, the GNI Principles address technology companies broadly defined, including Internet companies such as Meta, telecommunication companies and telecom equipment vendors. Moreover, GNI's work in the area, including its policy briefs, has adopted a broader perspective. As shall be outlined in the next chapter, transparent rule-making is one of the recurring standards in civil society instruments which require the development of content moderation standards to be open and participatory.

A recent GNI policy brief, in this regard, provides that the requirement of 'legality' requires that restrictions on free expression should be based on a clear and accessible law that is adopted through democratic legislative processes, particularly when the law-making powers of states are delegated (Global Network Initiative Policy Brief 2020, 12–13). Additionally, the policy brief draws from the requirement of 'legality' that the delegation of regulatory or adjudicatory roles to private companies must be accompanied by corollary safeguards of independent and impartial oversight. The policy brief further relates the transparency of content moderation practices, including the need for human review of content moderation practices, to the requirement of legality (Global Network Initiative Policy Brief 2020, 14). We consider in Sect. 2.4 to what extent such interpretive exercises help in reimagining international law to the content governance context. But the fact that GNI principles apply only to a dozen technology companies on a voluntary basis may lessen their impact.

---

[6] Note, though, that this is on top of an emerging body of scholarship that emphasises the need for authoritative interpretation by experts of the general and State-centred norms to make them fit for purpose. See, for example, Douek 2021; Benesch 2020a: 86, 86–91; Benesch 2020b: 13–14, 16–17.

[7] See details at https://globalnetworkinitiative.org/about-gni/.

In addition to freedom of expression, content governance engages a broad range of human rights guaranteed in international law that are yet to receive due attention in the Internet governance discourse. This is the other variant of content governance norms in human rights law. Common acts of content moderation such as content curation, flagging and take downs normally restrict freedom of expression of users. But other human rights and principles such as the right to equality/non-discrimination, the right to effective remedy, the right to fair hearing and freedom of religion would also be impacted by platform content moderation policies and practices. As in freedom of expression, the corresponding duties to these rights fall on states but they form the basis for recent civil society content governance standards which—in contrast—are addressed, in most cases, directly to both states and social media companies. This will be considered further in the next chapter.

As highlighted above, a key part of states' positive human rights obligation is to ensure effective remedy when violation of rights occurs (General Comment 31 2004, para 8). The right to effective remedy of individuals is corollary to that duty which entitles them to seek remedy when any of the Covenant rights, including freedom of expression, are violated. Thus, this is essentially a 'supporting guarantee', as opposed to a freestanding right (Joseph and Castan 2013, 869, 882). Content moderation would also engage cross-cutting rights such as the right to non-discrimination (International Covenant on Civil and Political Rights 1966, arts 2(1), 3, 26). Speech that incites discrimination or hatred against particular groups would violate the right to non-discrimination. Content moderation decisions to remove certain content might amount to violation of free speech rights while a decision to retain the problematic content might equally violate the right to equality/non-discrimination.

The right to a fair hearing guarantees fair processes to individuals in civil as well as criminal cases (Universal Declaration of Human Rights 1948, art 10; International Covenant on Civil and Political Rights 1966, art 14). This right imposes a duty to 'respect' on states, but its aim is to ensure respect for due process guarantees such as the ability to challenge charges through a fair and impartial process. According to the Human Rights Committee (HR Committee)—a treaty body that oversees the ICCPR, this right "serves as a procedural means to safeguard the rule of law" (General Comment 32 2007, para 1). Content governance decisions inherently give rise to due process concerns, for instance in the context of notification of decisions to users or in the opportunity to challenge those decisions.

Freedom to manifest religion would also be engaged by content governance. This right is the 'active' component of freedom of religion that entitles believers to freely express and practise their faith in any means, including through the use of digital platforms (Universal Declaration of Human Rights 1948, art 18; International Covenant on Civil and Political Rights 1966, art 18; Nowak 2005, 413–418). But this right is not absolute and hence may be restricted in line with the three-part requirements of legality, necessity and legitimacy. While the duty to respect and protect this freedom falls on states, content moderation decisions against content relating to the manifestation of religion or belief would constitute interference with the freedom to manifest religion and should be justified under the three-part tests.

Another relevant human right to be impacted by content moderation decisions is the protection of honour and reputation (International Covenant on Civil and Political Rights 1966, art 17). Honour relates to the subjective opinion of a person about oneself whereas reputation concerns the opinion of others about the person (Volio 1981, 198 et seq). The right is among the bundle of personality rights guaranteed in international law alongside the right to privacy (Universal Declaration of Human Rights 1948, art 12; International Covenant on Civil and Political Rights 1966, art 17). Decisions either to takedown, moderate or retain content in social media platforms that attack the honour or reputation of individuals would engage this right. As highlighted above, one of the legitimate aims for permissible restriction of freedom of expression is for ensuring respect of the rights or 'reputation of others'—and not honour (International Covenant on Civil and Political Rights 1966, art 19(3(a))). This is also the approach taken in the community guidelines of several digital platforms. A good case in point is Twitch's Terms of Service which prohibits defamatory content on its platform (Twitch 2021).

A less known but potentially relevant international standard concerns the right to a 'social and international order' in which human rights and freedoms (set forth in the UDHR) could be fully realised. Article 28 of the UDHR reads as follows:

> Everyone is entitled to a social and international order in which the rights and freedoms set forth in this Declaration can be fully realized.

This right is essentially aspirational in the sense that it requires "social and international conditions to be restructured" so as to enable the realisation of rights (Eide 1999, 597). According to Eide, this would mean readjustment of political and economic relations within states (social order) and among states (international order). The right does not envisage a clear corresponding duty or a duty-bearer. In light of the fact that the UDHR is a soft law, this is not surprising. But there can be no doubt that states would be the prime duty-bearers under the right to a rights-friendly social and international order. But as a right whose drafters had hoped would help create an order where rights, including those highlighted above could be realised, several actors—including social media companies as well as states—arguably bear responsibility, if not a duty, under this provision (Schabas 2013, 2753 et seq). Many scholars have alluded to the advent of a new social order with the rise of big technology companies (Zuboff 2018).[8] Together with states, these companies are responsible for ensuring that their practices in this new order—including on content governance—do not impair the enjoyment of rights.

Among the bundle of cultural rights guaranteed in international law is what has come to be referred to as the 'right to science'. Initially recognised in the UDHR, it guarantees the right of 'everyone' to share in "scientific advancement and its benefits" (Universal Declaration of Human Rights 1948, art 27(1)). It is later codified in the socio-economic and cultural rights covenant (International Covenant on Economic, Social and Cultural Rights 1966, art 15(1(b))). Article 15 of the International Covenant on Economic, Social and Cultural Rights (ICESCR) provides as follows:

1. The States Parties to the present Covenant recognize the right of everyone:

    (a) [...]
    (b) To enjoy the benefits of scientific progress and its applications.
    (c) [...]

2. The steps to be taken by the State Parties to the present Covenant to achieve the full realization of this right shall include those necessary for the conservation, the development and the diffusion of science and culture.

---

[8] Zuboff argues that surveillance capitalism is dominating the social order at the expense of freedom and democracy.

The nature of the right is such that it seeks to enable all persons who have not taken part in scientific progresses or innovations to participate in enjoying the benefits (Adalsteinsson and Thörballson 1999, 575–578). In that sense, it has the objective of protecting the rights of both scientists and the general public. This provision has barely been invoked in practice, but it might arguably apply to counter aggressive content moderation practices of social media companies vis-à-vis copyrighted materials. Digital platform policies routinely layout circumstances by which copyright infringing material may be subject to content moderation actions (TikTok 2021). As will be shown in the next chapter, 'freedom from censorship' is one of the content moderation-related standards often proposed in civil society initiatives. An aspect of this civil society-advocated freedom is the right to not to be subjected to onerous copyright restrictions.

More generally, the specific content of this 'right to science'—and the attendant obligations—is uncertain, however. As a second generation right, the realisation of the right to science is progressive. That means states owe no immediate obligations. But one of the progressive state obligations relevant to the question at hand is to take the necessary steps towards the diffusion of scientific outputs. According to an interpretation by the UN Special Rapporteur in the field of Cultural Rights, the right involves two key sub-rights (Special Rapporteur in the field of Cultural Rights 2012, paras 25, 43–44; General Comment 25 2020, para 74). The first sub-right concerns the right of individuals to be protected from the adverse effects of scientific progress. The other dimension of the right relates to the right to public participation in decision-making about science and its uses.

While the ensuing human rights obligations fall on states, these rights appear to resemble civil society content governance standards.[9] In the context of platform content moderation, the first sub-right would—for instance—require measures to prevent harm and safeguard social groups, including vulnerable groups, on social media platforms. The second sub-right might concern meaningful participation in the development of moderation policies. Likewise, this dimension of the right to science finds parallel in recent attempts by various actors, including civil society groups, to define the human rights responsibilities of digital platforms. We will return to this point in the next chapter.

In summary, a number of rights in international law potentially provide high-level normative guidance to content governance. But the fact that

---

[9] For more, see the next chapter.

these standards are generic in formulation—yet to be unpacked by authoritative bodies—means that this potential is unlikely to find meaningful practical application. Further punctuating this limitation are the complexities that the state-centred nature of the standards engenders. The UN Guiding Principles on Business and Human Rights not only are slightly specific in formulation but also address corporations more directly in human rights language. The next section explores this point further, particularly whether the UN business and human rights framework is fit for the purpose of digital content governance.

The UN framework on business and human rights, also referred to as the Ruggie Principles—named after its drafter John Ruggie, the late Special Representative of the UN Secretary General for Business and Human Rights—is the second potential source of specific international content governance standards. The Ruggie Principles is currently the only international instrument that seeks to address the conduct of businesses and the attendant impact on human rights (Guiding Principles on Business and Human Rights 2011). But it primarily affirms states as the sole and primary duty-bearers in human rights law, and hence it does not introduce new obligations (Guiding Principles on Business and Human Rights 2011, part I). This means, in turn, businesses, including technology companies, bear no human rights duties in international law. In human rights law—as highlighted at the outset, states' human rights obligations are of two types: negative and positive obligations. State positive obligation concerns the duty to ensure that human rights are not violated by third parties, including companies. Beyond elaborating this duty of the State, the UNGPs also introduces corporate human rights 'responsibilities' of businesses.

Structurally, the UNGPs is organised around three key normative pillars. First, it reaffirms and slightly unpacks states' human rights duty to protect in the specific context of human rights violations by businesses (Guiding Principles on Business and Human Rights 2011, part I, paras 1–10). It does so by requiring states to prevent, investigate, punish and redress abuse committed by businesses by putting in place the requisite legal and regulatory framework. But the duty to 'protect' would also apply in instances where states co-own businesses or otherwise deal with businesses whose conduct raises human rights concerns. Second, the Ruggie Principles imposes a 'responsibility' to respect on businesses (Guiding

Principles on Business and Human Rights 2011, part II, paras 11–24). Three points are worth noting regarding this aspect of the UNGPs.

One is the terminological choice. While states owe obligations or duties, businesses bear merely corporate responsibilities resulting in no legal consequences but simply moral obligations. Not complying with the Principles would not, thus, amount to a violation of international law but merely ignoring global expectations (Oliva 2020, 616). The other is that unlike for states, the responsibility is only to 'respect'—that is, a negative responsibility—and hence, businesses are not expected to 'protect' human rights. As part of the corporate responsibility to respect, businesses are 'expected' not to cause or contribute to human rights violations—and where they occur, to address the ensuing impact. To meet this responsibility to respect, businesses are expected to take the following steps: (a) to make policy commitment to respect human rights in their operations and dealings, (b) to undertake due diligence to prevent human rights violations and (c) to put in place processes of remediation.

Thirdly, the UNGPs envisages standards for the provision of remedies both by states and by businesses (Guiding Principles on Business and Human Rights 2011, part III, paras 25–31). When it comes to states, it stipulates that the duty to 'protect' embodies the obligation to put in place avenues for remediation by victims of human rights violations. And such ways of remediation could be either state or non-state based. Businesses, on the other hand, are expected to institute 'operational-level' mechanisms of handling grievances. Indeed, the Ruggie Principles also encourages other ways of remediation through industry and multistakeholder initiatives (Guiding Principles on Business and Human Rights 2011, part III, para 30).

The adoption of the Ruggie Principles is, no doubt, a significant development in terms of addressing corporations in human rights parlance. But neither its development nor its application thus far has focused on technology companies, including digital platforms. We return to this point in Sect. 2.4, but what immediately follows discusses how the emergent standards being introduced by intergovernmental mandates on freedom of expression address technology companies more directly. As shall be shown, this signifies further progress in the international law discourse relating to digital content governance.

3.3   EMERGENT PROGRESSIVE STANDARDS

Relatively progressive content moderation-related international standards are emerging through the work of UN special mandates, particularly the former UN Special Rapporteur on Freedom of Expression. In a series of reports the former Special Rapporteur, Kaye, examined—as highlighted above—the extent to which and whether international human rights law offers normative guidance in addressing the impact of content moderation practices on freedom of expression. But his role as part of the coalition of intergovernmental mandates on freedom of expression has had more significance in terms of outlining more progressive as well as normatively strong standards in the area of content moderation. Since 1999, intergovernmental mandates on freedom of expression—including the former UN Special Rapporteur—have issued 'Joint Declarations' on various themes relating to freedom of expression and media freedom.[10]

The declarations constitute international soft law that tends to unpack general free speech standards envisaged in the ICCPR—the main international hard law that guarantees the right to freedom of expression—as well as regional human rights treaties such as the European Convention on Human Rights and Fundamental Freedoms (ECHR) and the African Charter on Human and Peoples' Rights. The 2019 Declaration, for instance, states in its Preamble that the prime aim of the Joint Declarations is one of *"interpreting human rights guarantees* thereby *providing guidance* to governments, civil society organisations, legal professionals, journalists, media outlets, academics and the business sector" (emphasis added) (Joint Declaration 2019, preamble, para 3). It further provides that the Joint Declarations have—over the years—"contributed to the establishment of *authoritative standards"* on various aspects of free speech (emphasis added) (Joint Declaration 2019, preamble, para 4).

Intermediary liability is one of the content governance-related themes addressed at length in the joint declarations. As alluded to above, current international law standards on free speech are couched in general terms and offer little guidance to the complex and granular nature of content moderation practices. Apart from the right to freedom of expression guaranteed in the ICCPR and subsequent specialised human rights treaties, no international law instrument addresses the specific aspects of free speech protection in the digital environment. In particular, the role of

---

[10] See a list of the Joint Declarations at https://www.osce.org/fom/66176.

intermediaries such as social media platforms in curating and moderating content online is not addressed in international hard law. Intermediaries play a key role in the enjoyment of the right to freedom of expression online which makes appropriate regulation of their conduct an imperative. If intermediaries were to play an active role in moderating content circulating on their platforms to avoid liability, the free speech rights and interests of users would be seriously curtailed. The objective of a fair intermediary liability regime is, then, to define the exceptional circumstances where intermediaries would be held liable for the problematic content of their users. In offering standards on intermediary liability, the Joint Declaration fills—to an extent—the normative void in international law.

Subject to narrow exceptions, international human rights law—as alluded to above—imposes no direct duty on non-state actors, including corporations. But the joint declarations send mixed signals. For instance, the 2019 Joint Declaration states in its preamble that intermediaries such as social media companies owe a responsibility—not a duty—to 'respect' human rights. In so saying—and in line with the Ruggie Principles, it absolves corporations from any human rights duty. But in its operative provisions, the Declaration appears to address intermediaries more directly. As a soft law, it cannot introduce binding obligations, but it tends to use compulsory terms when addressing intermediaries. Paragraph 1(d) of a Joint Declaration adopted in 2017 provides the general principle of intermediary liability as follows:

> Intermediaries should never be liable for any third-party content relating to those services unless they specifically intervene in that content or refuse to obey an order adopted in accordance with due process guarantees by an independent, impartial, authoritative oversight body (such as a court) to remove it and they have the technical capacity to do that.

This general principle is further elaborated through specific principles. One such principle requires intermediaries to put in place clear and predetermined content moderation policies that are developed based on consultation with users (Joint Declaration 2017, para 4(a)). And such rules must set out objectively justifiable criteria that are not driven by political or ideological goals. Part of this requirement is that content moderation policies of intermediaries, including modalities of its enforcement, should be easily intelligible and accessible to users (Joint Declaration 2017, para 4(b)). This principle appears to reflect what is called in international human rights law the requirement of 'legality'. The other specific principle stipulates that intermediaries should institute minimum due process guarantees subject only to reasonable legal or practical constraints (Joint Declaration 2017, para 4(c)). This is mainly in two respects. One is that they should provide prompt 'notification' to users whose content may be subjected to 'content action' such as take down. Secondly, it requires intermediaries to put in place avenues by which users may challenge impending content actions. In the latter case, the Declaration mandates the need to ensure the coherence and consistency of content moderation decisions. The Declaration requires intermediaries to apply these standards to any automated (e.g. algorithmic) content moderation processes, but it permits exemption for "legitimate competitive or operational needs" (Joint Declaration 2017, para 4(d)). This reinforces the exemption that intermediaries may decline to enforce court take down orders when the measures are not technically feasible.

Joint declarations adopted in the following years reinforce the above standards in particular contexts. The 2020 Declaration, for instance, addresses the role of online intermediaries in relation to elections (Joint Declaration 2020, arts 1(c(iv)), 2(a(ii))). The 2021 Joint Declaration addresses freedom of expression and political actors where a series of recommendations are offered for social media companies. Among others, it calls upon social media platforms to ensure that their content moderation rules, systems and practices are clear and consistent with international human rights standards (Joint Declaration 2021, art 4). Particular focus is given in this Declaration to political advertisements. Social media platforms are called upon to adopt rules governing political advertisements which should (a) be clear and non-discriminatory, (b) be labelled as such, (c) indicate the identity of the sponsor, (d) enable users to opt-out of targeting and (e) be archived for future reference. The 2022 Joint Declaration deals with issues at the intersection of gender justice and freedom of expression. More relevant to the question at hand is that the Declaration calls upon digital platforms to ensure that content moderation policies and automated systems do not discriminate on the basis of gender or amplify and sustain gender stereotypes (Joint Declaration 2022, arts 1(e), arts 4(e), 5(c)).

Notably, the above highlighted norms introduce progressive standards on content governance, partly influenced by the work of civil society groups, including Internet bills of rights. Soft law generally offers authoritative interpretation of high-level principles of international hard law, but the approach in the joint declarations raises questions of form and substance in international law. One such question is whether a soft human rights instrument drawing upon a human rights treaty could directly address non-state actors that are not party to the underlying treaty. Related to this is the question of whether reading binding obligations in general binding instruments through progressive interpretation of soft law is tenable. Stated differently, the legal status of soft law of such form would ultimately undercut its normative value unless, of course, intermediaries choose to follow it regardless.

Another avenue by which such elaborative soft laws may earn more or better legal authority is if—for instance, the HR Committee—were to draw upon the declarations. That way, progressive standards provided in the joint declarations would get more audience and jurisprudential value. Yet interpretive bodies such as the Committee are shackled by structural constraints that make it hard for them to engage in elaborate content governance jurisprudence. We address this particular point in the next section.

3.4 REGULATORY LIMITS

To what extent does international law offer much sought-after normative guidance to platform governance? This section seeks to address this question in light of the brief sketch of the normative sources of international platform governance law in the preceding section. First, we consider the ways in which the relevant rules are designed to undercut the potential of international law in offering normative guidance on the governance of and in digital platforms. Second, we consider structural challenges that further punctuate the design constraints, namely the lack of robust oversight and accountability mechanisms.

3.4.1 Design Constraints

Design constraints of the international law of platform governance relate to the inherent normative characteristics of human rights standards more broadly. By design, international human rights law is state-centred. Only states are directly involved in its making and ultimately are obliged to respect and protect human rights. Non-state actors such as digital platforms may in one way or another play a role in shaping the making of rules of international law. But they are not subject to human rights obligations.

International law delegates to national law when it comes to regulating the conduct of digital platforms.

International human rights treaties are rarely universally ratified, which means that there will be states with no underlying human rights obligation to put in place the requisite legal and regulatory framework applicable to digital platforms. At the time of writing, the ICCPR has 173 state parties while the ICESCR has been ratified by 171 states.[11] A much lesser level of ratification has been recorded so far for the protocols of both Covenants that envisage an individual communications procedure. This leaves out dozens of states which remain with no obligation to legislate on content governance. Of course, treaty ratification is no guarantee for the existence of a robust domestic regulation. Many states might not be *willing* or *able* to follow upon their human rights commitments. Indeed, recent regulatory initiatives in some jurisdictions appear to be prompted more by jurisdiction-specific considerations than by human rights commitments, a good case in point being the recent deluge of 'fake news' legislation in many African countries (Garbe et al. 2021).

A related design constraint is that the relevant international standards are formulated in an exceedingly generic manner. Generic formulation of norms is generally desirable in making rules apply across time and rapidly changing technological environments. But the particularly truncated nature of international law standards relating to content governance undercuts their potential. Regulation of content in digital platforms requires rules that attend to the complexities and dynamism of the digital platform ecosystem. Generic international human rights standards are not suited to this reality of digital platforms. We will return to the point of how structural problems further punctuate this design constraint in the next section. But in the meantime, it suffices to note that institutional arrangements in the international human rights regime offer weaker oversight mechanisms that are beset further by structural problems. This means that there is little jurisprudence that would shed light on generically formulated standards.

It has been suggested that generic international law standards are sufficiently unpacked in the jurisprudence of regional and national courts (Kaye 2019, 42; Aswad 2018, 26, 58–59). But as proponents of adapting international law standards to content moderation acknowledge, not all requirements of Article 19(3) can readily be transposed. For instance,

---

[11] See details at https://indicators.ohchr.org/.

corporations—unlike states—cannot invoke 'public/national security' or 'public health' as a legitimate aim when restricting speech on their platforms (Lwin 2020, 69–70: Aswad 2020, 657–658). But they do in practice somehow. One recalls here the suspension of the Facebook and Instagram accounts of former US President Donald Trump which essentially invoked public security/safety as a legitimate aim for the decision (Zuckerberg 2021). In the wake of the current pandemic, social media companies have likewise updated their policies to attend to health-related speech. But the question of whether platforms can invoke such objectives as legitimate aims normally reserved to states remains.

A version of the design constraint relates to the applicability of the UN business and human rights framework to digital platforms. The development of the UNGPs did not originally have technology businesses in sight. Surprisingly so in light of the time when it was adopted—as recently as 2011. Nor has the work of the Working Group that oversees the Ruggie Principles in the past decade considered technology companies. Transnational corporations operating in mining and petroleum industries, among other business sectors, that pose tangible, brick-and-mortar human rights concerns were, and remain to be, the prime concern. For instance, a 2016 report of the Working Group, a body that oversees the Guidelines, focuses on the "human rights impact of agroindustry operations, particularly the production of palm oil and sugarcane, on indigenous peoples and local communities" (Report of the Working Group 2016).

An earlier report of the Working Group from 2014 even indicated that its areas of priority for the future will be promoting the incorporation of the Guiding Principles in the policy framework of 'international institutions' (Report of the Working Group 2014, para 84; Report of the Working Group 2021). A recent 'stocktaking' report of the Working Group explicitly acknowledges the hitherto exclusive focus on brick-and-mortar corporations and signals a shift towards technology companies in the future (Report of the Working Group 2021, paras 66, 74). This might gradually go some way in bringing technology companies within the radar of the Working Group. In this respect, the 2022 report of the UN Office of High Commissioner for Human Rights (OHCHR) charts the path where the ways in which the Ruggie Principles may apply to the technology companies are, to a degree, mapped (Report of the High Commissioner for Human Rights 2022). By way of a side note, it is vital to flag that the preparation and content of this report was informed by input from experts and the work of stakeholders from different geographic regions. This reinforces the increasing cross-fertilisation of norms between civil society initiatives and international human rights standards.

The inherently generic formulation of the Ruggie Principles further limits its potential of being applied to digital platforms. Indeed, the scope of the Ruggie Principles is defined in a manner to apply to businesses of all types and sectors (Guiding Principles on Business and Human Rights 2011, part II, para 14). This means that the principles would, theoretically, apply to technology companies, including social media platforms. But how, in practice, it would apply to them is uncertain, particularly because of the generic formulation of its principles. Attempts to adapt the principles to the world of content moderation—alluded to above—may gradually help refashion the content governance policies and practices of digital platforms. Yet, this avenue is not only rife with uncertainties, but it hinges very much on the good will of platforms and their readiness to heed to civil society advocacy.

Exacerbating the uncertainty is the resultant discrepant approach across platforms. Such uneven platform policies and practices would inevitably affect the rights of users. More certain would have been authoritative guidance from adjudicative bodies such as the HR Committee. But as the next section shall illustrate, such bodies operate in a framework crippled by structural constraints.

3.4.2   Structural Constraints

Structural constraints of the international law of content governance relate to the lack of effective institutional arrangements that would translate or apply generic and state-centred standards to the unique realities of the digital platform ecosystem. A characteristic feature of the international human rights system is that it relies on human rights bodies which operate in a framework that does not permit the development of elaborate and dynamic jurisprudence. Ad hoc oversight bodies organised into committees and working groups are the only international mechanisms of human rights accountability. But the committees and working groups are composed of experts that work part-time. For instance, the HR Committee, which is responsible for overseeing the ICCPR, is a body of 18 part-time experts who meet only three times per year, and only for four weeks each time (International Covenant on Civil and Political Rights 1966, art 28). Moreover, these part-time experts are expected to review state periodic reports (and adopt concluding observations), examine individual communications, develop general comments and hear reports of rapporteurs on the follow-up of views and concluding observations. Add to that the frequent reshuffle of members of the Committee, due to term limits and other factors, which affects the development of a coherent rights jurisprudence (Yilma 2023, Chap. 2).

Perhaps in an attempt to fill this void, special procedures of the Human Rights Council have taken steps to translate generic content governance standards to the platform context. In particular, several reports of the former UN Special Rapporteur on Freedom of Expression and Opinion, Kaye, have issued instructive thematic reports. Kaye often argues that the Guiding Principles provide 'baseline approaches', a useful 'starting point' that all technology companies should adopt (Report of the Special Rapporteur on Freedom of Expression and Opinion 2018a, para 70; Report of the Special Rapporteur on Freedom of Expression and Opinion 2016, paras 10, 14).[12] The UNGPs, according to Kaye, provides a 'global standard of expected conduct' of social media companies (Report of the Special Rapporteur on Freedom of Expression and Opinion 2018b, para 21). A recent report of the OHCHR likewise claims that the Ruggie Principles is an "authoritative global standard for preventing and addressing human rights harms connected to business activity, including in the technology sector" (Report of the High Commissioner for Human Rights 2022, paras 7–8). Its authority and legitimacy, the OHCHR further claims, flow from the fact that the instrument was endorsed by the Human Rights Council with wide private sector support and participation.

Kaye elaborates on how free speech principles should apply in specific contexts. With respect to hate speech in platforms for instance, he argues that social media companies should have an ongoing process to determine how hate speech affects human rights, institute mechanisms of drawing upon the input from stakeholders, including potentially affected groups, regularly evaluate the effectiveness of their measures, subject their policies to external review for the sake of transparency and train their content policy teams and moderators on human rights norms (Report of the Special Rapporteur on Freedom of Opinion and Expression 2019, paras 44–45). His report on Artificial Intelligence (AI) similarly reads into Ruggie Principles specific responsibilities of social media platforms.

---

[12] Note that in line with this trend globally, the Joint Declarations also appear to assert the applicability of the Ruggie Principles to social media companies (e.g. see Joint Declaration 2019, preamble, para 13, art 3(c)).

Among other points, it states that platforms should make high-level policy commitments to respect the human rights of users in all AI applications, avoid causing or contributing to adverse human rights impacts through the use of AI technologies, conduct due diligence on AI systems to identify and address potential human rights concerns, conduct ongoing review of AI-related activities, including through consultations and provide accessible remedies when human rights harms are caused by the use of AI technologies (Report of the Special Rapporteur on Freedom of Expression and Opinion 2018b, para 21).

Nevertheless, such expansive interpretation of rather crude human rights and principles can be problematic. Particularly, the problem relates to the normative authority of the interpretive exercises as well as the relevant legal instruments. The Ruggie Principles is, for instance, a soft law carrying no binding obligations. That UNGPs are inherently non-binding means that adherence by businesses is on a voluntary basis. Even if more states were willing and able to regulate businesses in their jurisdictions, the outcome would be an uneven level of protection among states. And this is undesirable in Internet regulation, which inherently involves transnational issues, including digital rights protection. Neither the joint declaration nor reports of UN special rapporteurs carry the type of authority needed to ensure compliance with international content governance standards, including the Ruggie Principles. The reports provide ways in which they would apply to the unique and specific context of social media companies. But the non-binding nature of such reports means that acceptance by platforms would be entirely voluntary. In international law-making, such reports of UN special rapporteurs do not count as standard-setting instruments but one that offer intellectual guidance on specific human rights standards. But they might carry some legal effect in offering authoritative interpretation of treaty provisions or other soft law instruments in a manner suited to particular contexts such as the unique contexts of technology companies.

Such an elaborative role helps provide normative guidance to companies as well as states. But as even advocates of adapting international law standards, including the Ruggie Principles, to the social media context acknowledge, the scope and content of corporate human rights responsibilities is in the process of development in international law (McGregor et al. 2020, 326). Added to frequent exhortations of civil society groups on the normative value of the Ruggie Principles, emerging attempts at translating the UNGPs to the digital context would contribute towards a crystallisation and further development of international standards on content governance. But this only means that the topic is in a state of flux which, in turn, makes it less suited to the exigencies of platform governance.

GNI's independent assessment of technology companies is a form of institutional oversight. Member technology companies undergo periodic assessment of their performance against GNI Principles. But it comes with structural limitations of its own. One is that the membership of technology companies is remarkably small. At the time of writing, 13 such companies are members of the initiative, and only a few of them—namely, Facebook and Google—are actively involved in content governance.[13] Another shortcoming is that the assessment is undertaken only every two or three years. That makes it less responsive to the dynamism in the digital ecosystem. But more crucially, as a voluntary oversight scheme, what the GNI Board—based on an independent assessment—could issue is a determination on whether the companies have made "good-faith efforts to implement the GNI Principles with improvement over time". The latest determination of the Board, for instance, has been in the affirmative (Global Network Initiative 2019). These appear to limit the value of the independent assessment mechanism in holding technology companies to account.

In the past few years, two developments that seek to increase the normative value of the UNGPs, including in extending its applicability to the technology sector, have emerged. One relates to attempts to crystallise the Ruggie Principles into hard law—that is, a treaty. But the draft treaty being negotiated by states has little to offer when it comes to technology companies and human rights (Third Draft Business and Human Rights Treaty 2021). Not only that the draft treaty does not pay specific attention to technology companies that wield enormous powers in the digital age but also that it does not even address businesses directly. Perhaps a notable innovation of the draft treaty is that it introduces a committee that would oversee the implementation of the treaty, fashioned like human rights treaty bodies. But what awaits the proposed Committee are the same structural setbacks that international institutional arrangements face. Operating part-time and meeting a few times a year, the future Committee would not probably be able to make international law fit for purpose to the digital platform ecosystem.

---

[13] See details at https://globalnetworkinitiative.org/#home-menu.

A manifestation of the structural challenges is the general comment of the HR Committee on freedom of expression, which provides no meaningful guidance to topical issues of content governance (General Comment 34 2011). The Comment offers an authoritative interpretation of Article 19 of the ICCPR. Although it was adopted as recently as 2011, it does not address content governance themes. Issues of content governance are unlikely to gain prominence before the HR Committee because of its 'slow-moving' jurisprudence as well its part-time roles meeting only thrice a year. A further reflection of the 'slow-moving' nature of UN jurisprudence generally on content governance is that recent resolutions on freedom of expression pay no apparent attention to the issues of content moderation. A good case in point is the series of resolutions adopted by the Human Rights Council under the label "Promotion, Protection and Enjoyment of Human Rights on the Internet". While these resolutions—adopted intermittently since 2009—focus primarily on the challenges of upholding freedom of expression on the Internet, topical issues of platform content moderation practices receive no mention (HRC Res 47/16 2021). A more recent Resolution 'affirm' the dicta that the same rights that people have offline must also be protected online. But content moderation by digital platforms is not addressed in any of the resolutions in meaningful detail. And understandably so, given the truncated nature of resolutions.

3.5   FILLING A VOID

This chapter explored the extent to which and whether the recent turn to international human rights law for content governance standards is a worthwhile exercise. It has shown that, beset by a host of design and structural constraints, international law does not offer meaningful normative guidance to the governance of and in digital platforms. A closer look at the current catalogue of international norms reveals that it remains uncertain just how international law would apply to the fast-moving, complex and voluminous nature of platform content governance. If major social media platforms such as Facebook follow through on their commitment to base their content moderation policies and practices in international human rights law, it would be along with all its uncertainties. In the absence of international oversight mechanisms, those companies are left with ample room to determine what international standards would apply, when and how when carrying out routine content moderation measures. Meta's inaugural human rights report highlights how the company's policies and practices are 'informed' by, 'drew' from and 'build' on international human rights standards (Meta 2022). With their obvious business interests in mind, this state of affairs diminishes the potential of international human rights law standards in providing reliable normative guidance on platform content governance.

A remarkable development, however, is that there are signs of progressive articulation and development of international standards into civil society instruments on content governance, a theme explored at length in the next chapter. While international law provides the overarching framework, civil society instruments appear to offer progressive standards on content governance at two levels. At one level, civil society standards are addressed directly to private actors, including social media companies, as well as states. This departs starkly from the state-centred international standards. But in doing so, civil society standards often tend to adapt state-centred international standards to social media companies. In that sense, there is a level of convergence between the two sources of content governance standards. At another level, civil society standards offer relatively detailed normative guidance on content governance. But such elaborate standards find only some form of high-level articulation in international law.

With the adoption of progressive standards in joint declarations, the normative cross-influence between international law standards and civil society standards is increasingly taking a new shape. But this phenomenon raises the question of what role international law should, or could, have in platform governance. In grounding progressive content governance standards that revitalise generic and state-centred international law standards, civil society initiatives tend to fill the void in international law. Nevertheless, the growing normative progression would mean little in practice unless the standards find proper articulation and recognition in international law and are advocated actively by civil society groups to influence the policies and practices of digital platforms. The next chapter shows how civil society initiatives are seeking to reimagine the international law of content governance.

References

Adalsteinsson, Ragnar, and Påll Thörballson. 1999. Article 27. In *The Universal Declaration of Human Rights: A Common Standard of Achievement*, ed. Guðmundur Alfreðsson and Asbjørn Eide. The Hague/Boston/London: Martinus Nijhoff Publishers.

Aswad, Evelyn. 2018. The Future of Freedom of Expression Online. *Duke Technology and Law Review* 17.

———. 2020. To Protect Freedom of Expression, Why Not Steal Victory from the Jaws of Defeat? *Washington & Lee Law Review* 77 (1).

Benesch, Susan. 2020a. But Facebook's Not a Country: How to Interpret Human Rights Law for Social Media Companies. *Yale Journal on Regulation Bulletin* 38.

———. 2020b. Proposal for Improved Regulation of Harmful Content Online. In *Reducing Online Hate Speech: Recommendations for Social Media Companies and Internet Intermediaries*, ed. Yuval Shany. Jerusalem: Israel Democracy Institute.

Committee on Economic, Social and Cultural Rights. 2020. General Comment No. 25: On Science and Economic, Social and Cultural Rights, UN Doc E/C.12/GC/25.

Douek, Evelyn. 2021. The Limits of International Law in Content Moderation. *UCI Journal of International* 6.

Eide, Asbjørn. 1999. Article 28. In *The Universal Declaration of Human Rights: A Common Standard of Achievement*, ed. Guðmundur Alfreðsson and Asbjørn Eide. The Hague/Boston/London: Martinus Nijhoff Publishers.

Garbe, Lisa, Lisa-Marie Selvik, and Pauline Lemaire. 2021. How African Countries Respond to Fake News and Hate Speech. *Information, Communication & Society*.

Germany. 2021. Network Enforcement Act (Netzdurchsetzunggesetz, NetzDG). Federal Law Gazette I.

Global Network Initiative. 2008, as updated in 2017. *GNI Principles on Freedom of Expression and Privacy*.

———. 2019. *GNI Principles at Work: Public Report on the Third Cycle of Independent Assessments of GNI Company Members*.

———. 2020. *Content Regulation and Human Rights: Analysis and Recommendations: Policy Brief*.

Henkin, Louis. 1999. The Universal Declaration at 50 and the Challenges of Global Markets. *Brooklyn Journal of International Law* 25 (1).

HR Committee. 2004. *General Comment No. 31: The Nature of the General Legal Obligation Imposed on States Parties to the Covenant*. UN Doc CCPR/C/21/Rev.1/Add.13.

———. 2007. *General Comment No. 32: Article 14—Right to Equality before Courts and Tribunals and to a Fair Trial*. UN Doc CCPR/C/GC/32.

———. 2011. *General Comment No. 34: Article 19—Freedoms of Opinion and Expression.* UN Doc CCPR/C/GC/34.

Joseph, Sarah, and Melissa Castan. 2013. *The International Covenant on Civil and Political Rights: Cases, Materials and Commentary.* Oxford: Oxford University Press.

Karavias, Markos. 2013. *Corporate Obligations Under International Law.* Oxford: Oxford University Press.

Kaye, David. 2019. *Speech Police: The Struggle to Govern the Internet.* New York: Columbia Global Reports.

Land, Molly. 2019. Regulating Private Harms Online: Content Regulation under Human Rights Law. In *Human Rights in the Age of Platforms*, ed. Rikke Jørgensen. Cambridge, MA: MIT Press.

Lwin, Michael. 2020. Applying International Human Rights Law for Use by Facebook. *Yale Journal on Regulation Online Bulletin* 38.

McGregor, Lorna, Daragh Murray, and Vivian Ng. 2020. International Human Rights Law as a Framework for Algorithmic Accountability. *International Comparative Law Quarterly* 68 (2).

Meta. 2021. Corporate Human Rights Policy. https://bit.ly/3WAOiNg.

———. 2022. Meta Human Rights Report 2020–2021: Insights and Actions. https://bit.ly/3Z8pb6a.

Nowak, Manfred. 2005. *U.N. Covenant on Civil and Political Rights: CCPR Commentary.* Kehl am Rhein: Engel.

Oliva, Thiago. 2020. Content Moderation Technologies: Applying Human Rights Standards to Protect Freedom of Expression. *Human Rights Law Review* 20 (4).

Opsahl, Torkel, and Vojin Dimitrijevic. 1999. Articles 29 and 30. In *The Universal Declaration of Human Rights: A Common Standard of Achievement*, ed. Guðmundur Alfreðsson and Asbjørn Eide. The Hauge/Boston/London: Martinus Nijhoff Publishers.

Ratner, Steven. 2001. Corporations and Human Rights: A Theory of Legal Responsibility. *The Yale Law Journal* 111 (3).

Schabas, William, ed. 2013. *The Universal Declaration of Human Rights: The Travaux Préparatoires— Vol. III.* Cambridge, UK: Cambridge University Press.

TikTok. 2021. Intellectual Property Policy. https://bit.ly/3G8guAa.

Twitch. 2021. Terms of Service. https://bit.ly/3CdopLw.

Twitter. 2022. Defending and Respecting the Rights of People using our Service. https://bit.ly/3G7baNl.

United Nations. 1948. *Universal Declaration of Human Rights.* UN General Assembly Resolution 217 A (III).

———. 1965. *International Convention on the Elimination of All Forms of Racial Discrimination.* UN General Assembly Resolution 2106 (XX).

———. 1966. *International Covenant on Civil and Political Rights.* UN General Assembly Resolution 2200A (XXI).

————. 1966. *International Covenant on Economic, Social and Cultural Rights.* UN General Assembly Resolution 2200 (XXI).

————. 1989. *Convention on the Rights of the Child.* UN General Assembly Resolution 44/25.

————. 2007. *Convention on the Rights of Persons with Disabilities.* UN General Assembly Resolution 61/106.

————. 2011. *Guiding Principles on Business and Human Rights: Implementing the United Nations 'Protect, Respect and Remedy' Framework.* UN Doc A/HRC/17/31.

————. 2012. *Report of the Special Rapporteur in the field of Cultural Rights, Farida Shaheed, on the Right to Enjoy the Benefits of Scientific Progress and Its Applications.* UN Doc A/HRC/20/26.

————. 2014. *Report of the Working Group on the Issue of Human Rights and Transnational Corporations and Other Business Enterprises.* UN Doc A/HRC/26/25.

————. 2016a. *Report of UN Special Rapporteur on Freedom of Expression and Opinion, David Kaye.* UN Doc A/HRC/32/38.

————. 2016. *Report of the Working Group on the Issue of Human Rights and Transnational Corporations and Other Business Enterprises.* UN Doc A/71/291.

————. 2018a. *Report of UN Special Rapporteur on Freedom of Expression and Opinion, David Kaye.* UN Doc A/HRC/38/35.

————. 2018b. *Report of UN Special Rapporteur on Freedom of Expression and Opinion, David Kaye.* UN Doc A/73/348.

————. 2019. *Report of UN Special Rapporteur on Freedom of Expression and Opinion, David Kaye.* UN Doc A/47/486.

————. 2021. *Guiding Principles on Business and Human Rights at 10: Taking Stock of the First Decade, Report of the Working Group on the Issue of Human Rights and Transnational Corporations and Other Business Enterprises.* UN Doc A/HRC/47/39.

————. 2021. *Legally Binding Instrument to Regulate, in International Human Rights Law, the Activities of Transnational Corporations and Other Business Enterprises, Third Revised Draft.* https://bit.ly/3i2yTD1.

————. 2021. *Report on the Seventh Session of the Open-ended Intergovernmental Working Group on Transnational Corporations and other Business Enterprises with respect to Human Rights.* UN Doc A/HRC/49/65.

————. 2021. *The Guiding Principles on Business and Human Rights: Guidance on Ensuring Respect for Human Rights Defenders, Report of the Working Group on the Issue of Human Rights and Transnational Corporations and other Business Enterprises.* UN Doc A/HRC/47/39/Add.2.

————. 2021. *The Promotion, Protection and Enjoyment of Human Rights on the Internet.* HRC Res 47/16.

————. 2022. *The Practical Application of the Guiding Principles on Business and Human Rights to the Activities of Technology Companies, Report of the Office of the United Nations High Commissioner for Human Rights.* UN Doc A/HRC/50/56.

United Nations, Organization for Security and Cooperation in Europe, Organization of American States and African Union. 2017. Joint Declaration on Freedom of Expression and 'Fake News', Disinformation and Propaganda.

————. 2019. Twentieth Anniversary Joint Declaration: Challenges to Freedom of Expression in the Next Decade.

————. 2020. Joint Declaration on Freedom of Expression and Elections in the Digital Age.

————. 2021. Joint Declaration on Politicians and Public Officials and Freedom of Expression.

————. 2022. Joint Declaration on Freedom of Expression and Gender Justice.

United States Supreme Court. 2019. *Manhattan Community Access Corp. v. Halleck*, 139 S. Ct. 1921, 1930.

Volio, Fernando. 1981. Legal Personality, Privacy and Family. In *The International Bill of Rights: The Covenant on Civil and Political Rights*, ed. Louis Henkin. New York: Columbia University Press.

Yilma, Kinfe. 2023. *Privacy and the Role of International Law in the Digital Age.* Oxford: Oxford University Press.

Zuboff, Shoshana. 2018. *The Age of Surveillance Capitalism: The Fight for a Human Future at the New Frontier of Power.* New York: Public Affairs.

Zuckerberg, Mark. 2021. Facebook Post on the Suspension of Donald Trump. https://bit.ly/3vCpM2M.

CHAPTER 4

Shaping Standards from Below: Insights from Civil Society

Abstract This chapter discusses how civil society groups are articulating rights and principles for the digital age through non-legally binding declarations, known as 'Internet Bills of Rights'. These documents represent the 'voice' of communities that are seeking to redefine core constitutional principles in light of the challenges posed by digital society, resulting in a new form of 'digital constitutionalism'. The chapter analyses 40 Internet Bills of Rights and their principles related to online content governance, including substantive and procedural standards, as well as ad hoc provisions specifically crafted to address social media platforms. These provisions aim to contextualise and adapt international human rights standards into more granular norms and rules to be implemented in the platform environment.

Keywords Digital constitutionalism • Platform governance • Internet Bill of Rights • Content governance • Content moderation

4.1 A Constitutional 'Voice' and 'Bridge'

The rise and diffusion of social media has generated novel communicative spaces blurring the boundaries between public and private. On the one hand, social media represent inherently private spaces insofar as they are owned by private companies, and interactions within them are mainly

E. Celeste et al., *The Content Governance Dilemma*, Information Technology and Global Governance,
https://doi.org/10.1007/978-3-031-32924-1_4

regulated through private governance means such as contracts, terms of service, community standards or internal policies. On the other hand, some structural characteristics of social media, from the ease of access to interactivity and horizontal flow of communications, led many scholars to conceive them as "an infrastructure capable of revitalizing and extending the public sphere" (Santaniello et al. 2016) after its downfall as an effect of consumerism and the rise of mass media as depicted by Habermas (1992). Some researchers saw social networks as 'third places' that, like Habermasian 'coffee shops' in eighteenth century England, serve as a new, easily accessible forum for public life, promoting social interactions and political debate (Chadwick 2009; Farrell 2012).

Leaving aside the question of to what extent social media play a positive role in democracy by fostering participation, civic engagement and people's empowerment against political and elite structures (Bimber et al. 2012), or rather they put at risk the democratic process favouring manipulation, extremism and polarisation (O'Connor and Weatherall 2019; Benkler et al. 2018), there is no question that they are increasingly relevant in forming the public opinion. Consequently, social media companies' rules end up shaping the limit of what can be considered an acceptable exercise of the freedom of speech for billions of people carrying out de facto an intrinsically public function (Jorgensen and Zuleta 2020; Celeste 2021a). Additionally, social media platforms have the ability to blur the boundaries of the dichotomy between the public and private dimensions. As transnational companies, these platforms facilitate cross-border communication and contribute to softening frontiers and demarcations within and outside nation states as well as between jurisdictions and territories, thus making sovereignty claims more complex and uncertain (Celeste 2021b; Celeste and Fabbrini 2020). As Grimm (2016) pointed out, this twofold erosion of the state authority caused by transnational modes of governance brings a serious challenge to the constitutional order and guarantees. Constitutionalism, in its traditional sense, requires the "concentration and monopoly of public power that allows a comprehensive regulation" on a territory and the identification of a polity, acting as "pouvoir constituent", and establishing forms of self-limitation in the exercise of public power (Santaniello et al. 2018).

The "constitutionalisation" of international law (De Wet 2006) has been proposed as the remedy to the pitfalls caused by transnational models of governance. This perspective, rather than looking for a "legitimatory monism and an unilateral form of law-production by a political subject"

(Moller 2004, 335), focuses on "continuity, legitimatory pluralism and the spontaneous evolution of a legal order" (Idem). In such a view, some norms of international law may fulfil constitutional functions and then acquire a constitutional quality, integrating and verticalising the international order (Gardbaum 2008; De Wet 2006), in a kind of "compensatory constitutionalism" (Peters 2006) that completes and fills the gaps created by globalisation in domestic constitutional systems (Santaniello et al. 2018; Celeste 2022a).

Even if theories about the constitutionalisation of international law identify interesting tendencies and solutions to counterbalance the erosion of nation-state authority, nevertheless, they have little to say about how to safeguard constitutional guarantees and fundamental rights within transnational private (or mainly private) regimes carrying out public functions. International Human Rights Law is far from constitutionalising the international order, and as seen in the previous chapter, due to its state-centred design, it does not directly impact private actors that own and rule social media platforms. Furthermore, its generic formulation of principles and norms appears unfit to regulate a complex socio-technical environment such as platform content moderation which requires a rather granular and dynamic system of rules.

Then, the recent proliferation of civil society initiatives advocating for human rights standards on social media platforms may not be a mere coincidence. These efforts are part of a larger movement to articulate rights and principles for the digital age. The output of these initiatives often takes the form of non-binding declarations, intentionally adopting a constitutional tone and thus referred to as "Internet bills of rights" (Celeste 2022b). These declarations can be seen as expressions of the *voice* of communities that are seeking to redefine core constitutional principles in light of the challenges posed by digital society, resulting in a new form of "digital constitutionalism" (Santaniello and Palladino 2022). The growing number of civil society digital constitutionalism initiatives in the content governance field could be conceived as a reaction, on the one side, to the increasing power of social media platforms in shaping public opinion, and, on the other side, to the impracticability to directly apply international human rights law standard within platforms' transnational private governance regime. These efforts may be seen as an attempt to *bridge* constitutional thinking with the everyday governance of social media platforms (Palladino 2021b).

In this regard, Gunther Teubner's theory of societal constitutionalism can provide a sound conceptual framework to understand how civil society's bills of rights can play this role. Based on Luhmann's theory of social systems (1975) and the subsequent developments by Sciulli (1992) and Thornhill (2011), the German scholar moves his considerations starting from the dynamics of social differentiation. From this point of view, the more a social subsystem becomes autonomous, the more it develops 'its own systemic logic based on a specific means of communication' that makes possible and meaningful the interaction within the subsystem (such as the money in the economic subsystem and the law for the legal subsystem). As the activities of a subsystem become relevant to the social system as a whole, they give raise to what Teubner calls "expansionist" and "totalizing" tendencies (Teubner 2011, 2012), meaning that the subsystem can impose its logic on the other social spheres to reproduce itself, threatening the integrity and autonomy of individuals and communities.

According to this perspective, the rise of the Internet and digital technologies in our societies can be conceived as a process of autonomisation of an emerging digital subsystem. In the wake of Lessig (2006), we can identify in the code the communicative means of the digital subsystem, meaning by this not some programming language, but rather the socio-technical architecture which, by combining software, hardware and human components, makes the interaction between different social actors in the digital world possible, shaping their experience and disciplining their behaviour. While the code constitutes the means of communication of the digital subsystem, digitisation or datafication (George 2019) can be interpreted as its logic. The latter therefore consists of an incessant process of conversion of social reality into digital information in order to be further processed and elaborated by systems to extract new information with added value.

The constitutionalisation of a subsystem occurs when frictions with other social spheres bring out "fundamental rights" understood as "social and legal counter-institutions" (Teubner 2011, 210). This allows, on the one hand, to free the "potential of highly specialized dynamics" of the subsystem, and on the other hand, to institutionalise self-limitation mechanisms that preserve the integrity and autonomy of individuals and other social spheres (Teubner 2004, 12). From this point of view, fundamental rights perform both an *inclusive* function, guaranteeing universal access to the specific 'means of communication' of the subsystem and therefore to the related rule-making processes, and an *exclusive* function, in the sense of defining the boundaries of the subsystem's sphere of action. A qualifying aspect of Teubner's theory consists in the idea that fundamental rights can be constituted within a social subsystem only through a process of generalisation and re-specification—which means that their functions, to be effective, must take place in the 'communication medium' of the subsystem and inscribed in its operating logic. Furthermore, Teubner's societal constitutionalism appears as a hybrid constitutionalisation process, in which the self-limitation of a subsystem is the result of the pressures, resistances and constraints posed by other social spheres.

These considerations indicate firstly, that for fundamental rights to be truly effective in the social media environment, they must be translated and incorporated into their socio-technical architecture, including programming, algorithms, internal policies and operational routines (Palladino 2021a, 2022). Secondly, they point out that limiting mechanisms for platforms cannot be based solely on forms of self-regulation nor on state regulation. Certainly, states can impose constraints on Big Tech, both through the means of ordinary legislation and by exercising a "shadow of hierarchy" on self-regulatory processes, threatening the imposition of heavy regulation on a sector if certain standards are not reached (Héritier and Lehmkuhl 2008). However, in order for these mechanisms to be effective and overcome the obstacles posed by the private, transnational and infrastructural nature of digital processes, they must be completed and accompanied by the joint action of a plurality of actors (Palladino 2021b).

Among the actors involved in this process of hybrid constitutionalisation, civil society organisations and their Internet Bills of Rights play a crucial role (Celeste 2019). In the first place, civil society carries out a fundamental 'watch-dog' function, documenting human rights violations by both states and corporations, giving a voice to common users, vulnerable groups and minorities, shedding light on the human rights implications of platform policies and functionalities, and new pieces of legislation. Secondly, civil society organisations can exert pressure on both states and companies to adopt proper instruments and mechanisms to comply with human rights standards, thus starting the above-mentioned process of generalisation and re-specification of fundamental rights for the digital world. Indeed, by drafting Internet Bills of Rights, NGOs and activists can draw on a consolidated corpus of norms and reflections elaborated in the international human rights law ecosystem and put them in the concrete context of social media platform reality. Moving from their expertise in human rights violations, civil society organisations could identify what kind of practices and operations need to be banned, fixed or introduced, defining the rules and operational practices for this purpose. Of course, this is just a first step in the process of translating human rights standards into the socio-technical architecture of platforms, which require further phases of elaboration by legislators, technical communities and platform owners themselves before becoming fully implemented arrangements. Nevertheless, it is a crucial step to foster state intervention and push companies to take into account their responsibilities, creating a convergence of expectations around a common normative framework. The more civil society organisations engage in global conversations and networking, the more likely it becomes for them to converge on a series of norms and practices for social media platforms. Insofar as it happens, civil society can facilitate the reach of a global standard, influencing both national legislation and companies' practices. Of course, this kind of outcome cannot be taken for granted. Differences in cultural and political backgrounds or the social context in which they operate may lead different human rights defenders to pay more attention to some specific issues rather than others, to conceptualise the same problems differently or to prefer alternative approaches.

This chapter investigates to what extent civil society's Internet Bills of Rights have been able to so far bridge international human rights law and platform governance, translating human rights standards into more granular norms for the social media platform environment. The examination will also consider the extent to which global civil society efforts converge on a shared normative framework, which has the potential to shape both state regulations and corporate policies and contribute to the development of a global standard.

4.2   Civil Society and Internet Bills of Rights

In order to investigate how civil society is contributing to the constitutionalisation of social media content governance, we performed a content analysis on a corpus of Internet Bills of Rights extracted from the Digital Constitutionalism Database. The Digital Constitutionalism Database is an online accessible and interactive resource resulting from the joint efforts of researchers taking part in the Digital Constitutionalism Network based at the Center for Advanced Internet Studies (Bochum, Germany).[1] The database collects more than 200 documents (Internet Bills of Rights; declarations of digital rights; resolutions, reports, policy briefs containing recommendations on digital rights), which are drafted by different kinds of actors (civil society organisations, parliaments, governments, international organisations, business companies, multistakeholder initiatives) from 1996 up to now, engaging with the broad theme of the exercise and limitation of power on the Internet and seeking to advance a set of rights, principles and governance norms for the digital society.

The Digital Constitutionalism Database has been analysed in order to select documents drafted by civil society groups discussing online content governance conceived as the set of rules and practices through which decisions are made about the hosting, distribution and display of user-generated content by Internet service providers. Since social media platform content moderation is a relatively recent issue, the broader concept of content governance has also been used in order to draw lessons from general principles coming from older documents and monitor trends over time.

A total of 40 documents were identified based on the established selection criteria. The geographic and temporal distributions of the selected documents are presented in Figs. 4.1 and 4.2, respectively. As shown, attention towards the relationship between content governance and digital rights has, not surprisingly, grown together with the rise of social media platforms from the second half of the 2000s. The majority of the documents in our corpus were generated by organisations that assert their transregional or global reach. These entities comprise coalitions of civil society groups from across the globe, such as the Association for Progressive Communications, and the Just Net Coalition, or individual civil society organisations that maintain offices in various continents with personnel and governing structures that reflect a variety of backgrounds, including ARTICLE 19, Access Now and Amnesty International. This circumstance may facilitate the emergence of a cohesive framework at the global level, given that these global civil society associations constitute an exercise in global networking that can synthesise diverse experiences, concerns and claims from various contexts. However, upon examination of the organisations that are more closely associated with a particular national or regional

---

[1] The Digital Constitutionalism Database is accessible online at: https://digitalconstitutionalism.org/database/.

Fig. 4.1 Geographical distribution of the analysed documents

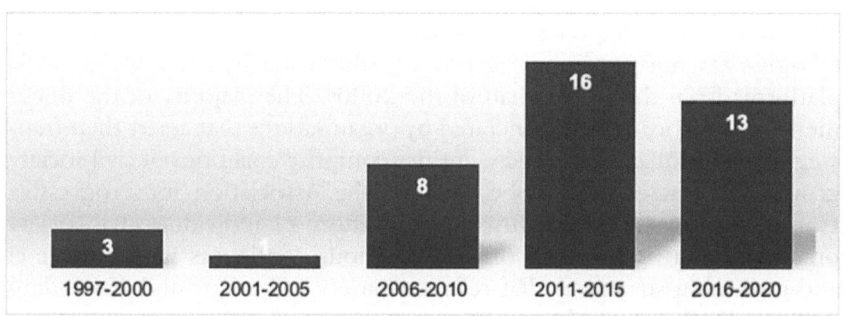

Fig. 4.2 Distribution over time of the analysed documents

background, we observe that very few cases in our corpus originate from Africa and Asia. The reasons for this may be attributed to challenges in collecting documents drafted in non-European languages, resource constraints of these organisations and obstacles faced when operating in non-democratic countries.

With regard to the content of the document in our corpus, constitutional and governance principles have been hand-coded with the NVIVO software resorting to an inductive methodology (Bazeley and Jackson

2013; Kaefer et al. 2015). In the first stage, principles have been coded as closely as possible as they appeared in the text. At a later stage, synonymous items have been merged as well as principles aggregated in a hierarchical system of broader categories. It is worth noting that this exercise of coding and categorisation faces an unavoidable degree of overlapping and redundancy. On the one hand, indeed, several principles detected in the texts could frame the issue in slightly different ways, or on the contrary, the same principle is employed highlighting different features of the same concept, or again, some principles could cover part of the semantic area of a broader one. On the other hand, the categories created to aggregate more close-to-text coding reflect the authors' interpretative framework and are settled to emphasise distinctions and aspects deemed to be relevant by researchers. Besides the limits of the qualitative approach, redundancy appears to be a characteristic feature of digital rights itself, since "these rights and principles are more often than not interconnected, interdependent, mutually reinforcing, and in some cases even in conflict with one another" (Gill et al. 2015).

Table 4.1 provides a synthetic overview of the over 90 principles we detected in the corpus, organised and summarised into broader categories. The first one collects all the provisions explicitly concerned with international human rights law compliance. The other two categories distinguish between substantive and procedural principles, drawing on the distinction between substantive and procedural law. In this context, 'substantive principles' refer to people's expected behaviour according to accepted social norms as well as their basic human rights such as life and liberty. In this case, more specifically, substantive standards for content governance indicate people's rights and responsibilities related to the creation and publication of online content. By contrast, 'procedural principles' indicate formal rules and procedures through which substantive rights are created, exercised and enforced. In this case, more specifically, procedural standards indicate the rules through which decisions about users' contents are made, including the rulemaking process itself (Main 2010; Alexander 1998; Grey 1977).

In the first instance, data seems to outline a common framework, suggesting a remarkable degree of consensus among our sample of civil society initiatives on a shared set of principles to be applied to content governance. Civil society initiatives analysed in the corpus strongly rely on human rights law. Half of our sample refers explicitly to one of the international human rights law instruments discussed in the previous section

Table 4.1 Civil society initiatives

Categories	No. of documents	Principles included
General compliance with human rights standards	19	
Substantive principles	**39**	
Freedom of expression	38	Freedom from censorship, freedom from copyright restriction, freedom of religion
Prevention of harm	16	Harassment, cyberbullying, defamation, incitement to violence, cybercrime, human dignity
Protection of social groups	13	Non-discrimination of marginalised groups, discriminating content, hate speech, children rights and protection
Public interest	6	Public health or morality, public order and national security, fake news and disinformation, protection of infrastructure layer
Intermediary liability	9	Full immunity, conditional liability, intermediaries are liable in the case of actual knowledge of infringing content, intermediaries are liable when failing to comply with adjudicatory order
Procedural principles	**32**	
Rule of law	24	Legality, legal certainty, rule of law, judicial oversight, legal remedy, necessity and proportionality
Good governance principles	19	Transparency, accountability, fairness, participation, multistakeholderism, democratic governance
Platform-specific principles	21	Notification procedures, human oversight, human rights due diligence, limitations to automated content moderation, informed consent, right to appeal and remedy

(especially ICCPR, UDHR, Ruggie principles), or more generally, claims for the respect of international human rights standards. However, even when not quoted explicitly, the documents we analysed refer to rights, principles and standards drawn from the international human rights literature. Almost all the civil society charters (39 out of 40) deal with some substantive principles, mostly freedom of expression (38 out of 40). Three other categories of substantive principles, namely 'prevention of harms', 'protection of social groups' and 'public interest', mostly set the borders of acceptable exceptions of freedom of expression that justify content removal. Moreover, 34 out of the 40 documents analysed, mention some procedural principles, in particular those related to the rule of law (24), good governance principles (19) or procedural principles specifically tailored for social media platforms (21). Taken as a whole, procedural principles specify a series of conditions and requirements to exercise content moderation in a legitimate and rightful manner.

The convergence of civil society around the same framework appears more evident if we look at trends over time. For this purpose, we grouped the detected principles into five categories: (1) freedom of expression; (2) freedom of expression limitations, including 'prevention of harm', 'protection of social groups' and public interest; (3) intermediary liability; (4) rule of law; (5) other procedural principles, for the most part related to social media platform governance. Figure 4.3 shows that, while freedom of expression consistently remains the primary concern of civil society, when competing issues arise that may potentially curtail freedom of expression (e.g. hate speech, discrimination and child protection), such concerns are typically accompanied by demands for procedural rules to govern content moderation and prevent its use as a tool for undue censorship. It is noteworthy that in recent years, there has been a shift in focus from requesting states to adopt a legal framework to establishing rules and procedures directly targeted at social media platforms and private companies. This last step seems to indicate that civil society organisations matured

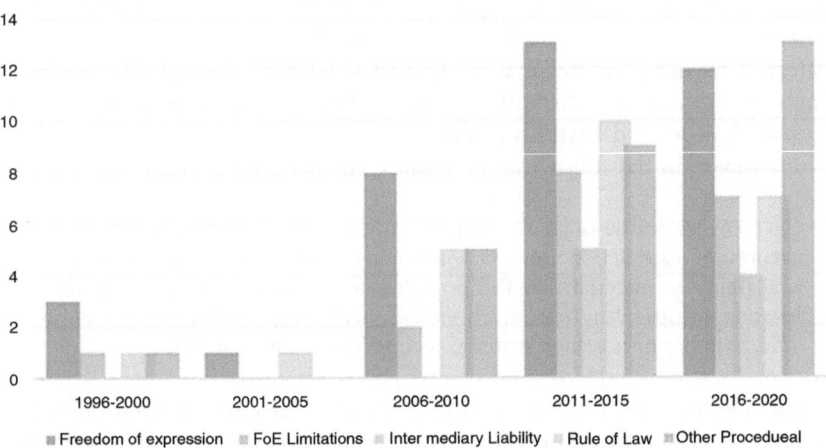

Fig. 4.3   Trends over time

enough experience and knowledge to start the process of translation of international human rights standards into a complex of norms and mechanisms to be embedded within the socio-technical architecture of social media platforms.

4.3   DEFINING SUBSTANTIVE RIGHTS AND OBLIGATIONS

4.3.1   Avoiding the Traps of Intermediary Liability

Substantive law constitutes the segment of the legal system that outlines the rights and obligations of individuals and organisations. It also pertains to the rules establishing the legal outcomes that arise when such rights and obligations are breached. A preliminary question to address when reasoning about substantive principles for social media content governance concerns the responsibilities of intermediaries and Internet service providers dealing with user-generated content. This question arose in the 1990s with the massification and commercialisation of the Internet and found a first answer in Section 230 of Title 47 of the 1996 US Telecommunication Act, which states that "no provider or user of an interactive computer service shall be treated as the publisher or speaker of any information provided by another information content provider". Section 230 conceived service providers not involved in content production as mere conduit or passive intermediaries, and therefore settled a different liability regime compared with traditional media carrying out editorial tasks. The decision aimed at safeguarding the then-nascent Internet service provider market against legal risks and external interferences that could have discouraged innovation and investment in this new sector (Bowers and Zittrain 2020).

However, in the following paragraph, the act affirms that:

> No provider or user of an interactive computer service shall be held liable on account of—any action voluntarily taken in good faith to restrict access to or availability of material that the provider or user considers to be obscene, lewd, lascivious, filthy, excessively violent, harassing, or otherwise objectionable, whether or not such material is constitutionally protected.

As Gillespie observed, in so doing the US legislator allowed providers to intervene "on the terms they choose, while proclaiming their neutrality as a way to avoid obligations they prefer not to meet", "without even being held to account as publishers, or for meeting any particular standards" of effective policing for how they do so, and ultimately ensuring them with "the most discretionary power" on content governance (Gillespie 2018, 30–33).

This "broad immunity" model has been a cornerstone for content moderation in the Western world. However, not even the US "broad immunity" model can be considered a full immunity, due to the existence of particular exceptions, most notably copyright infringements under the Digital Millennium Copyright Act (DMCA). Other Western countries seem to be converging towards a "conditional liability" model, largely influenced by the European Directive on e-Commerce,[2] according to which immunity is provided to intermediaries insofar as they have no "actual knowledge" of illegal content and they comply with the so-called notice and takedown procedure, timely removing illicit thirdparty content in compliance with state authorities or court requests (MacKinnon et al. 2015). In some contexts, notifications by users too are deemed to presume actual knowledge by intermediaries in case of "manifestly illegal content" (CoE 2018).

However, civil society organisations claim that conditional liability, in the absence of a formal legal framework specifying clear rules and safeguards, will not limit the degree of arbitrariness with which platforms police users' content, but rather, could lead to 'voluntary' proactive measures and over-removals. In particular, the recent German Network Enforcement Act, or NetzDG, has been particularly criticised for its overbroad definition of unlawful content and the disproportionate sanctions for platform administrators in case of non-compliance, resulting in a delegation of censorship responsibilities to social media companies induced to err on the side of caution and undermine due guarantees.[3] Over time, civil society has developed a nuanced understanding of intermediary liability, seeking to avoid both the traps of platforms' arbitrariness stemming from broad immunity and the incentivisation of unlawful over-removal associated with conditional liability regimes.

---

[2] Directive 2000/31/EC of the European Parliament and of the Council of 8 June 2000 on certain legal aspects of information society services, in particular electronic commerce, in the Internal Market ('Directive on electronic commerce').

[3] The Network Enforcement Act led several German civil society associations to draft a "Declaration on Freedom of Expression" explicitly "in response to the adoption of the Network Enforcement Act". Criticism to the NetzDG has been advanced also in Access Now (2020), APC (2018), and ARTICLE 19 (2018).

Internet Bills of Rights define what intermediaries' rights and obligations are, focusing on three main points. First, they reiterate the principle according to which "no one should be held liable for content on the Internet of which they are not the author" (African Declaration Coalition 2014) and intermediaries should then be protected by a safe harbour regime against states pressures to undertake censorship on their behalf by imposing, *de jure* or *de facto*, a general monitoring obligation. Second, they "oppose full immunity for intermediaries because it prevents them from holding any kind of responsibility, leaving victims of infringement with no support, access to justice, or appeal mechanisms" (Access Now 2020), and thus, they propose exceptions to the safe harbour regime for cases in which intermediaries fail to comply with a court or other adjudicatory body's order to remove content, or do not take any action after being properly notified about potential illegal and harmful content. In this regard, advocacy groups also proposed alternatives to the usual notice and takedown procedure. They suggested implementing a 'notice-wait-and-takedown' procedure, which would require intermediaries to forward notices about alleged harmful or illegal content to users, giving them the opportunity to modify or remove the content themselves or object to the notice before the content is removed. Another proposed alternative is a 'notice-and-notice' procedure, under which Internet service providers are only legally required to forward notifications to alleged infringers. Third, in order to limit the arbitrary nature of platforms' content moderation, even in the case the latter police users' content on their own initiative, civil society organisations stated that platforms should adhere to clear rules and standards grounded in international human rights law. This issue is addressed in further detail in the section below on procedural principles. At the same time, in a couple of cases (Access Now 2020; APC 2018), civil society organisations contested the assumption at the basis of the current intermediary liability regime, arguing that content curation practices through which platforms foster user engagement put into question their role as passive intermediaries. However, in the words of the Association for Progressive Communication, this does not mean making "platforms such as Facebook legally liable for the content carried on the platform, but there is a clear need for more transparency and accountability in how they manage and manipulate content and user data" (APC 2018).

4.3.2   The Centrality of Freedom of Expression

Reasoning about content governance in terms of substantive principles means questioning the fundamental values that must be promoted and protected when we communicate through social media platforms. In this regard, civil society organisations show a clear stance. They consistently prioritise freedom of expression as the primary concern when addressing issues related to content governance in most of the documents we analysed.[4] This is not surprising considering that, since the beginning, freedom of expression has been a cornerstone of any attempts to safeguard human rights and establish constitutional principles for the digital sphere (Gill et al. 2015; Kuleska 2008). Freedom of expression constitutes for civil society the 'lens' through which it is possible to frame content moderation, meaning that the other detected content moderation principles, for their vast majority, just set the boundaries of permissible limitations on freedom of expression. This approach to content moderation differs from the approach taken by many nation states and incorporated in most of the platform terms and conditions or community standards, which often prioritise the definition and prohibition of non-acceptable content and behaviours.[5]

The difference could appear trivial, but it is of crucial importance. In the former case, the focus is entirely on the protection of freedom of expression, which could be exceptionally constrained only under narrowly defined conditions grounded on international human rights law standards. In the latter case, the focus is instead on the content to be removed in the

---

[4] Of the two charters not mentioning freedom of expression, the Santa Clara Principles proposes procedural principles for content moderation implicitly referring to freedom of expression, and the iRights Charter focuses on the digital right of child and young people.

[5] Some governmental Internet Bill of Rights such as the Magna Carta for Philippine Internet Freedom and the Nigerian Digital Right and Freedom Bill devote great attention to defining illegal content and behaviours on the Internet, such as copyright infringement, child pornography, defamation or hate speech and related punishment. However, it is worth noting that this operation takes place within a broader framework inspired by international human rights law. As shown later, civil society organisations generally welcome these attempts to establish a clear and foreseeable legal framework inasmuch as it provides proper guarantees. National laws, disjointed from a digital rights discourse and very often resulting in the outsourcing of judiciary and enforcing functions to private actors on the basis of vague definitions of cybercrime and disproportionate sanctions, such as the German NetzDG, Pakistan's Electronic Crimes Act (PECA), or the Nigerian Hate Speech Bill, are deemed much more alarming and dangerous for freedom of expression by civil society organisations.

name of a "public health" interest, "establishing accountability for concrete harms arising from online content, even where addressing those harms would mean limiting speech" (Bowers and Zittrain 2020). Not by chance, in the civil society charters, statements on freedom of expression are often coupled with a call to international human rights standards' compliance. Furthermore, most of the time, freedom of expression is put at the centre of a human rights and democratic value system, as exemplified in this excerpt from the Principles on Freedom of Expression and Privacy drafted by the Global Network Initiative (GNI 2018):

> Freedom of opinion and expression is a human right and guarantor of human dignity. The right to freedom of opinion and expression includes the freedom to hold opinions without interference and to seek, receive and impart information and ideas through any media and regardless of frontiers. Freedom of opinion and expression supports an informed citizenry and is vital to ensuring public and private sector accountability. Broad public access to information and the freedom to create and communicate ideas are critical to the advancement of knowledge, economic opportunity and human potential.
>
> The right to freedom of expression should not be restricted by governments, except in narrowly defined circumstances based on internationally recognized laws or standards. These restrictions should be consistent with international human rights laws and standards, the rule of law and be necessary and proportionate for the relevant purpose.

4.3.3   Setting the Boundaries of Freedom of Expression

As previously mentioned, the other substantive principles outlined in the civil society charters largely define the limits of acceptable exceptions to freedom of expression, providing justification for the removal of specific content. The first set of principles refers to the protection of harm and has been detected in 16 documents. According to these texts, "certain very specific limitations to the right to freedom of expression may be undertaken on the grounds that they cause serious injury to the human rights of others" (IRPC 2010, 2015), as well as to their reputation and dignity, or when they "involve imminent danger to human beings" (EDRi 2014), including the cases of harassment, cyberbullying and incitement to violence. Some other principles could be gathered under the label of "protection of social groups", since they aim to protect vulnerable or marginalised groups and ensure the full enjoyment of their rights. This kind of principle has been coded in 13 documents.

From a conceptual perspective, the protection of particular social groups could be broken down into three different modalities. First, we have principles providing that Internet operators will not discriminate against content on the basis of users' race, colour, sex, language, religion or other status. This principle is particularly relevant in content moderation since platform policies and automated tools end up disproportionately impacting more vulnerable and marginalised groups (APC 2018). A second group of principles aims to guarantee a safe environment for particularly vulnerable groups, especially children, which should be protected from exploitation and troubling or upsetting scenarios online. Third, we have principles calling to remove content which is discriminating or incites hostility and violence against minorities, vulnerable and marginalised groups, namely hate speech. A last cluster of cases (six documents) refers to restrictions of freedom of expression based on some ideas of public interest, often recalling the ICCPR, which mentions in this regard "the protection of national security or of public order, or of public health or morals". More recent civil society charters are also considering fake news and disinformation issues. It is worth noting that civil society organisations usually refer to the above-mentioned freedom of expression exceptions in very general terms. They appear reluctant to provide criteria identifying the cases requiring content moderation and call states to carry out this duty following human rights standards. As discussed in more detail in the next paragraph, once identified possible freedom of expression exceptions, civil society organisations right after specify that those exemptions must follow international human rights standards and procedural rules. Moreover, looking at their frequency, it seems that civil society is more likely to recognise legitimate freedom of expression exceptions when they deal with other individual rights, or behaviours capable of concretely impacting individual integrity and dignity, rather than in cases involving more abstract and collective values that could be more easily employed to allow undue censorship.

4.4   LIMITING PLATFORMS' ARBITRARINESS THROUGH PROCEDURAL PRINCIPLES

4.4.1   A Rule of Law Regime

From the perspective of civil society, procedural principles have a crucial relevance because they guarantee that substantive rights competing with freedom of expression, whatever they may be, are not misused or abused resulting in undue censorship practices affecting people's fundamental rights. Most of the civil society efforts to provide social media content governance with procedural safeguards and guarantees could be collected under the "rule of law" label. It would be overly simplistic to define the rule of law as a single principle. It can be understood as a multifaceted concept encompassing both a political philosophy and a series of mechanisms and practices, aiming at preventing the arbitrary exercise of power by subordinating it to well-defined and established rules and affirming the equality before the law of all members of a political community, including and foremost, the decision-makers (Walker 1988; Choi 2019). The rule of law "comprises a number of principles of a formal and procedural character, addressing the way in which a community is governed" (Waldron 2020), and which entail basic requirements about the characteristics of law, how it should be created and enforced. Laws should be accessible to all, general in form, and universal in application. Legal standards should be stable and legal responsibilities should not be imposed retrospectively. Moreover, laws should be internally consistent and provide for legal mechanisms to solve eventual conflicts between different norms.

The rule of law also implies the institutional separation between those who establish and enforce the law. Laws should be created or modified according to pre-established rules and procedures by bodies that are representative of those that will be affected by them. Furthermore, law should be applied impartially by independent judicatory bodies. According to Article 14 ICCPR, everyone charged with a criminal offence shall be entitled to due process and fair trial, entailing minimum guarantees, including among others "to be informed promptly and in detail in a language which he understands of the nature and cause of the charge against him; to have adequate time and facilities for the preparation of his defence and to communicate with counsel of his own choosing; to be tried without undue delay". Finally, public decisions and the law itself must be subject to judicial review to ensure that decision-makers are acting in accordance with the law, first and foremost constitutional and human rights law.

Civil society attempts to establish a 'rule of law' regime for Internet content governance focused on the request to establish a proper legal framework. The adoption of an accessible legal framework, indeed, with clear and precise rules, provides both Internet intermediaries and online users with legal certainty and predictability, ensuring that everyone is fully aware of their obligations and rights and is able to regulate their conduct properly. Above all, a sound legal framework is also a guarantee against the eventuality that constitutional safeguards are circumvented by outsourcing online content moderation adjudication and enforcement to private entities through opaque and non-human-rights-compliant terms of service, secretive agreements or codes of conduct. Furthermore, civil society groups particularly claim that states must not impose a 'general monitoring obligation' to intermediaries, conceived as a "mandate to undertake active monitoring of the content and information that users share [...] applied indiscriminately and for an unlimited period of time" (Access Now 2020, 24). Human rights defenders fear that encouraging a 'proactive' content moderation will lead to "over-removal of content or outright censorship" (APC 2018; ARTICLE 19 2017; EDRi 2014). In doing so, civil society groups recall the warnings advanced by the UN Special Rapporteur, David Kaye, in his 2018 Report on the promotion and protection of the right to freedom of opinion and expression (Kaye 2018).

According to civil society organisations, the legal framework for content moderation should at its minimum:

1. Provide clear definition of harmful and illegal content and of the conditions under which freedom of expression could be limited by law, through democratic processes and according to international human rights law standards.
2. Clearly establish under which conditions intermediaries are deemed responsible for user-generated content, and which kind of actions they must undertake. This also includes the conditions according to which an intermediary is supposed to acquire 'actual knowledge' of any infringing content. A legal framework should clarify which different duties, obligations and procedures stem from court orders, government requests, private notifications and flagging.

3. Encompass the content removal procedures including the timeframe for the different phases of the process; the obligation to notify users about content takedown.
4. Guarantee appropriate judicial oversight over content removal and the right to legal remedy, including the obligation to notify users about content takedown and provide them with all the necessary information to object against the removal decision.

The demand for the establishment by states of a legal framework for content moderation corresponds to a mirror request addressed to social media companies not to proceed with removing content unless prescribed by law, and in any case, trying to protect and promote human rights. For example, where requested by government to take actions that may result in a violation of human rights, companies should "interpret government demands as narrowly as possible, seek clarification of the scope and legal foundation for such demands, require a court order before meeting government requests, and communicate transparently with users about risks and compliance with government demands" (African Declaration Coalition 2014).

Another commonly referred procedural principle (mentioned in 15 papers) is the test of necessity and proportionality, which, together with the prescription by law and the pursuing of legitimate aim, is part of the international human rights standards for permissible freedom of expression limitations. According to ARTICLE 19, necessity requires "to demonstrate in a specific and individualised fashion the precise nature of the threat to a legitimate aim, [...] in particular by establishing a direct and immediate connection between the expression and the threat identified", while proportionality means that "the least restrictive measure capable of achieving a given legitimate objective should be imposed" (ARTICLE 19 2018).

4.4.2   Good Governance Principles

Besides the rule of law, civil society organisations proposed other procedural principles, which do not necessarily relate to the legal system itself, but which could be considered 'good governance' principles.

Transparency is recalled in 16 charters, and it is deemed crucial to achieve good content governance standards. According to the Association for Progressive Communication, "increased transparency is needed in a number of areas in order to better safeguard freedom of expression against arbitrary content removals and to better understand how the content viewed online is being moderated" (ARTICLE 19, 2018), while Access Now points out that "transparency is a precondition for gathering evidence about the implementation and the impact of existing laws. It enables legislators and judiciaries to understand the regulatory field better and to learn from past mistakes" (Access Now 2020). A consideration that could also be extended to private policies. If the adoption of an accessible legal framework is considered a basic transparency requirement for states, similarly, private companies are called to make their internal content moderation rules and procedures public in order to make content decisions predictable and understandable to users. Especially in the case of automated systems of content moderation and curation, full transparency is required in order to allow independent assessment, monitoring and evaluation.

Furthermore, civil society requires companies to timely provide users with all the information about the content moderation process in which they are involved. Both states and companies are required to report about content removal activities in a regular and public manner. Governments are asked to disclose information about all their requests to intermediaries that result in restrictions of freedom of expression. Companies are called to publish data about content removal, including both those following governmental requests and their own terms of services. Moreover, accountability is frequently mentioned among good governance principles. However, it tends to overlap with transparency, or appeal and remedy procedures, while it is almost totally lacking the reference to some kind of oversight mechanisms capable of reviewing platforms' rules and procedures against external independent bodies.

A discrete number of documents (13) call for participatory rule-making and decision-making in both public and private spheres. Some of them express this concept in very general terms, which include content governance even if not specifically tailored for this purpose. According to them, both public and private governance processes should be open, inclusive and accountable, allowing for the meaningful participation of everyone affected and "expand[ing] human rights to the fullest extent possible".[6] Among those who directly faced content moderation regulation, it is pos-

---

[6] African Declaration Coalition, 2014, African Declaration on Internet Rights and Freedom. On the same line, see also the NetMundial Statement.

sible to observe some differences. Some actors, such as EDRi and the "Community input on Christchurch call" place greater emphasis on "democratic" governance, meaning that responsibilities for speech regulation rely on democratically elected bodies, and they must not be outsourced to companies in order to ensure legal and constitutional safeguards. Some others, like Article 19, are worried that state intervention will pressure companies towards forms of over-removal, and are more favourable to self-governance arrangements, once provided that they are informed on international human rights standards and open to stakeholder participation.

4.5   Embedding Human Rights Standards into Platform Socio-Technical Design

In the last few years, civil society organisations seemed to move forward on the road of digital constitutionalism by contextualising and adapting the very general international human rights standards into more granular norms and rules to be implemented in the platform environment.

4.5.1   Transposing the Rule of Law

Most of the efforts of civil society in this regard have been devoted to generalising and respecifying a 'rule of law' regime for the social media platform context. In the first place, platforms are asked to provide a degree of certainty and predictability for their content moderation rules and procedures through accessible terms of service and community standards which is comparable with the one ensured by law in order to "to enable individuals to understand their implications and regulate their conduct accordingly" (Article 17). Content moderation rules and procedures must be publicly available, easily accessible, delivered in the official language of the users' country, and written in a plain language, avoiding obscure references to technical or legal jargon. However, social media companies should be transparent about the laws and regulations they follow, and they should inform individuals about situations in which the company may be required to comply with state requests or demands that could affect users' rights and freedoms (ARTICLE 19 2017; APC 2018).

An important element in the attempt to establish a 'rule of law' regime within the social media platform ecosystem relates to appeal and remedy procedures, which should reproduce some of the due process and fair trial rights' guarantees. It is worth noting that according to these civil society organisations, the establishment of these procedures does not prevent or limit users from resorting to traditional legal means; it rather introduces a further faster, more affordable and more immediate channel to claim their rights. According to the Santa Clara Principles (ACLU 2018), "companies should provide a meaningful opportunity for timely appeal of any content removal or account suspension, whose minimum standard includes: 'i) human review by a person or panel of persons that was not involved in the initial decision; ii) an opportunity to present additional information that will be considered in the review; iii) Notification of the results of the review, and a statement of the reasoning sufficient to allow the user to understand the decision'".

Companies are also requested to provide remedies, such as: "restoring eliminated content in case of an illegitimate or erroneous removal; providing a right to reply; with the same reach of the content that originated the complaint, offering an explanation of the measure; making information temporarily unavailable; providing notice to third parties; issuing apologies or corrections; providing economic compensation" (Access Now 2020). In particular, social media companies are expected to provide notice to each user whose content has been subject to moderation decisions, recalling well-known due process principles according to which courts cannot hear a case unless the interested party has been given proper notice. This notification must include at its minimum:

1. The indication of the alleged harmful or illegal content, which must be made accessible to the content provider by reporting it in the notification entirely, or including at least relevant excerpts, or by providing the URL or other information allowing for its localisation.
2. The specific clause of the terms of service, guidelines, community standards or law that has been allegedly violated. If the content has been removed as a result of a legal order or at the request of a public authority, the notification should also include the allegedly infringed law, the issuing authority and the identifier of the related act.
3. Details about the methods used to detect and remove the content, such as user flags, government reports, trusted flaggers, automated systems or external legal complaints.
4. An explanation of the content provider's rights and the procedures for appealing the decision or seeking legal review and redress.

Notification should go along with the ability for content providers to revise their posts in order to prevent or overcome content removal decisions, and to submit a counter-notification when they believe that their content was removed in error, explaining their reasoning and requesting that the content be restored. Counter-notifications should be considered a key element of the appeal procedures and a broader 'right to defence' in the content moderation context.

However, as stated in the previous section, the rule of law also implies that rules are established or modified through a democratic decision-making process, in turn, shaped by well-defined and pre-established rules. Furthermore, according to the rule of law, there should be some kind of institutional separation between those who create, execute and adjudicate rules and decisions. None of this exists in the social media platform environment. Content moderation rules are typically created by social media platforms' legal and policy teams, which are accountable to their top management and shareholders rather than to the affected communities, and when experts and stakeholders are involved, this occurs in a mere consultive role. Internal policies could be easily modified or dismissed according to companies' interests and leadership views. Content removal decisions are taken by platforms' employees with no guarantee of independent judgement, appeal or review.

It appears evident that this degree of arbitrariness in content governance significantly undermines platforms' efforts to transpose a rule of law regime into their governance structure. For this reason, some civil society organisations proposed to create independent self-regulatory bodies entrusted with the duty to define content moderation criteria and oversee their application. In particular, Article 19 (2018) suggested establishing an ad hoc 'Social Media Council' following the example of previously successful experiences such as press councils. Its independence from any specific platform, as well as its accountability and representativeness, should be guaranteed by a multistakeholder governance structure. The Council should adopt a Charter of Ethics for social media consistent with international human rights standards; draft recommendations clarifying the interpretation and application of ethical standards; review platforms' decisions under the request of individual users with the faculty to impose sanctions in the case of unethical behaviour violating the Charter. As part of their membership within a Social Media Council, platforms would have to commit to making their content moderation practices auditable by the Council, provide it with economic resources on a long-term basis and accept

Council decisions as binding. This would help to ensure that the Council is able to effectively oversee and regulate the platform's content moderation practices and that the platform is accountable for its actions.

4.5.2 Human Rights by Design

At the beginning of this chapter, we stated that any attempt to constitutionalise online content governance, in order to be effective, needs to embed fundamental rights into the socio-technical design of social media platforms. In the last few years, civil society organisations are becoming increasingly aware of this need and, not by chance, they are asking more frequently that platforms adopt a human rights by design approach. This consists of incorporating human rights considerations into the design and development of a platform from the very beginning(or one of its tools, applications or other components), rather than trying to address fundamental rights violations and abuses at a later stage when the platform has already been launched and scaled-up (Access Now 2020; Reporters Sans Frontiers 2018; ARTICLE 19 2017). Besides developing clear and specific policies and guidelines for content moderation based on human rights standards, this approach also implies their integration into the platform's user experience and underlying technical infrastructure. It includes providing training and support to the platform's users and moderators on the use of the platform itself and its content moderation policies, monitoring their effectiveness and making adjustments as needed to ensure that they are achieving their intended goals.

Embedding human rights standards into platforms' socio-technical design means translating and implementing them into organisational arrangements, management systems and technical specifications. This task is entrusted to the platforms themselves and the technical community; however, civil society organisations can play a key role by pressuring for the adoption of a human rights by design approach and by monitoring the effectiveness of the implemented arrangements. Moreover, civil society can give a specific contribution to constitutionalising social media by developing and promoting the adoption of instruments such as human rights impact assessments and human rights due diligence (Access Now 2020; APC 2018; Reporters Sans Frontiers 2018), through which social media platforms can scrutinise on an ongoing basis their policies, products and services with the consultation of third-party human rights experts in order to evaluate their impact on human rights. Companies are also called to share information and data with researchers and civil society organisations and to support independent research. In so doing, organisations can gain a better understanding of the potential impacts of their practices on human rights, develop strategies for addressing any negative externality and ensure that their content moderation practices are consistent with their human rights obligations.

4.5.3   Automated Content Moderation

Most of the discussion on how to embed constitutional principles for content governance within the socio-technical infrastructure of social media platforms has been focused on the specific topic of automated content moderation. The latter could be defined as the employment of algorithms and artificial intelligence in order to "classify user-generated content based on either matching or prediction, leading to a decision and governance outcome (e.g. removal, geo-blocking, account takedown)" (Gorwa et al. 2020, 3). Although the use of automated moderation systems is seen as essential for addressing increasing public demands for social media platforms to take greater responsibility for the content on their platforms, it introduces a further dilemma for content governance. On the one hand, automated systems allow facing the scale and pace of communication flows on social media platforms. On the other hand, ensuring the rule of law and human rights requires that decisions must be taken on an individual basis according to a series of procedures and guarantees. Civil society's main concern here is that automated content moderation may result in "general monitoring" practices (Access Now 2020), raising serious human rights concerns, both for freedom of expression and for privacy. Additional concerns have been raised regarding the accuracy, fairness, transparency and accountability of the process, due to certain technical factors.

Automated content moderation systems could be distinguished between matching and classifying methods to identify harmful and illegal content. The formers generate a unique identifier, or 'hash', for each digital content uploaded on the platform and then compare them to a database of already known hashes, related, for example, to copyrighted materials, content ordered to be removed by a public authority or previously classified as harmful or illegal. The latter use machine-learning algorithms to analyse digital content in order to automatically detect and classify content that violates certain rules or policies. To this purpose, machine-learning algorithms are trained on large datasets of digital content that have been previously labelled as harmful or illegal. The algorithm would then use this training data to learn the characteristics of content that is likely to be inappropriate. Once the algorithm has been trained, it can be applied to new content to automatically classify it as appropriate or inappropriate. Both methods have proven to be unable to distinguish contextualised uses of language or language nuance, such as irony, sarcasm, contents reported to denounce their inappropriateness or on the contrary covert threat, leading to both systemic false positive and false negative classifications. Furthermore, civil society organisations pointed out that filtering techniques such as hash-matching, which remove content before they are uploaded, may deprive civil society, academics and law enforcement of a precious trove of evidence to identify and prosecute human rights abuses (APC 2018).

The most relevant issues, however, are posed by machine-learning classifications. One of the major challenges is the potential for biases at the various stages of the machine-learning pipeline. These biases can manifest in a number of ways, such as through the over-or underrepresentation of certain types of content in training datasets, the reflection of cultural, linguistic or political prejudices in labelling or the introduction of bias during the phases of data processing, feature engineering or model hyperparameter setting by developers. Additionally, external factors such as adversarial or poisoning attacks can also introduce bias into the system. These sources of error can lead to a systematic disparate impact that disproportionately affects certain social groups or types of content. Machine-learning content moderation also poses relevant issues in terms of accountability and transparency (Smith 2020; Pasquale 2015). Especially when deep learning algorithms are employed, understanding and explaining how moderation decisions are made may be challenging or impossible even for the same people who created them (Palladino 2022). Deep learning algorithms are complex and hierarchical, with multiple layers of interconnected nodes that process and analyse initial input data into more and more complex mathematical functions unintelligible to the human mind. Furthermore, these systems evolve and adapt over time as they analyse new cases undermining the possibility of providing certainty and predictability for content moderation decisions.

The aforementioned concerns raise doubts as to whether automated content moderation is compatible with the rule of law. The Global Forum for

Media Development, in its Statement on the Christchurch Call,[7] stated that automated content removal "cannot currently be done in a rights-respecting way" and advocated for the rejection of "unaccountable removal of content" and "incentives for over-removal of content" (Global Forum for Media Development 2019). Similarly, the Zeit foundation, in its Charter of Digital Fundamental Rights of The European Union, affirmed, "Everyone has the right not to be the subject of computerised decisions which have significant consequences for their lives", and added, "Decisions which have ethical implications or which set a precedent may only be taken by a person". Even when admitted, automated content governance should undergo specific limits and conditions. According to Access Now, "the use of automated measures should be accepted only in limited cases of manifestly illegal content that is not context-dependant, and should never be imposed as a legal obligation on platforms" (Access Now 2020, 26), for example, sexual abuses against minors, while in the other cases algorithms can be used to flag suspicious content but the final decision should be taken by human operators. However, one should consider the automation bias, which is the tendency for humans to over-rely on automated systems, even when they may not be the best decision-making tool.

In any case, individuals should be notified when automated systems are being utilised for the policing of their content, and they have to be afforded the opportunity to request a human review of such decisions. Furthermore, companies should be required to provide an explanation of the ways in which automated detection is used across different categories of content, as well as the reasoning behind any decisions to remove said content. By and large, civil society associations ask companies to tackle well-known accuracy, transparency, accountability and fairness raised by automated content governance by adopting a human rights by design approach putting constitutional standards at the centre of the design, deployment and implementation of artificial intelligence systems (Palladino 2021a, 2022). Automated systems should comply with transparency requirements, providing as far as possible accessible explanations on their functioning and the criteria employed for their decisions, as well as information about procedures behind and beyond the application, including appeal and remedy mechanisms.

[7] The Christchurch Call is an agreement that was reached in May 2019 between several countries and major technology companies with the goal of combating the spread of terrorism and violent extremism online. The agreement was named after the city of Christchurch, New Zealand, where a terrorist attack was carried out at two mosques in March 2019, resulting in the death of 51 people. The Christchurch Call includes a number of commitments from participating countries and companies, including the promotion of media literacy, the development of new technologies to help identify and remove extremist content, and the sharing of best practices for preventing the spread of terrorism and violent extremism online. For more details, see https://www.christchurchcall.com/.

According to the Santa Clara Principles, companies should ensure that their content moderation systems "work reliably and effectively", pursuing "accuracy and non-discrimination in detection methods, submitting to regular assessments", and "actively monitor the quality of their decision-making to assure high confidence levels, and are encouraged to publicly share data about the accuracy of their systems". Civil society organisations recommend that the quality and accuracy of automated content moderation systems must be assessed through third-party oversight and independent auditing. Therefore, such systems must be designed to allow such an external scrutiny by means of proper traceability measures and documentation.

REFERENCES

Alexander, L. 1998. Are Procedural Rights Derivative Substantive Rights? *Law and Philosophy* 17 (1): 19–42.

ACLU Foundation. 2018. The Santa Clara Principles on Transparency and Accountability in Content Moderation. https://santaclaraprinciples.org

ARTICLE 19. 2017. Universal Declaration of Digital Rights. https://www.article19.org/resources/internetofrights-creating-the-universal-declaration-of-digital-rights/

———. 2018. Self-regulation and 'hate speech' on social media platforms. http://europeanjournalists.org/mediaagainsthate/wp-content/uploads/2018/02/Self-regulation-and-%E2%80%98hate-speech%E2%80%99-on-social-media-platforms_final_digital.pdf

Bazeley, Pat, and Kristi Jackson. 2013. *Qualitative Data Analysis with NVivo.* 2nd ed. London: Sage.

Benkler, Y., R. Faris, and H. Roberts. 2018. *Network Propaganda: Manipulation, Disinformation, and Radicalization in American Politics.* Oxford University Press.

Bimber, B., A. Flanagin, and C. Stohl. 2012. *Collective Action in Organizations: Interaction and Engagement in an Era of Technological Change.* Cambridge: Cambridge University Press.

Bowers, John, and Jonathan Zittrain. 2020. Answering Impossible Questions: Content Governance in an Age of Disinformation. *The Harvard Kennedy School (HKS) Misinformation Review.* https://doi.org/10.37016/mr-2020-005.

Celeste, E. 2019. Terms of Service and Bills of Rights: New Mechanisms of Constitutionalisation in the Social Media Environment? *International Review of Law, Computers & Technology* 33 (2): 122–138. https://doi.org/10.1080/13600869.2018.1475898.

———. 2021a. Digital Punishment: Social Media Exclusion and the Constitutionalising Role of National Courts. *International Review of Law, Computers & Technology.* https://doi.org/10.1080/13600869.2021.1885106.

————. 2021b. Digital Sovereignty in the EU: Challenges and Future Perspectives. In *Data Protection Beyond Borders: Transatlantic Perspectives on Extraterritoriality and Sovereignty*, ed. F. Fabbrini, E. Celeste, and J. Quinn, 211–228. Hart.

————. 2022a. The Constitutionalisation of the Digital Ecosystem: Lessons from International Law. In *Digital Transformations in Public International Law*, ed. A. Golia Jr., M.C. Kettemann, and R. Kunz, 47–74. Nomos Verlagsgesellschaft mbH & Co. https://doi.org/10.5771/9783748931638-47.

————. 2022b. *Digital Constitutionalism: The Role of Internet Bills of Rights*. Routledge.

Celeste, E., and F. Fabbrini. 2020. Competing Jurisdictions: Data Privacy Across the Borders. In *Data Privacy and Trust in Cloud Computing*, ed. G. Fox, T. Lynn, and L. van der Werff. Palgrave Macmillan.

Chadwick, A. 2009. Web 2.0: New Challenges for the Study of e-Democracy in an Era of Informational Exuberance. *I/S: A Journal of Law and Policy for the Information Society* 5 (1): 11–41.

Choi, N. 2019. Rule of Law. *Encyclopedia Britannica*.

COE. 2018. Recommendation CM/Rec(2018)2 of the Committee of Ministers to Member States on the Roles and Responsibilities of Internet Intermediaries.

De Wet, E. 2006. The International Constitutional Order. *International & Comparative Law Quarterly* 55 (1): 51–76.

EDRi. 2014. The Charter of Digital Rights. https://edri.org/wp-content/uploads/2014/06/EDRi_DigitalRightsCharter_web.pdf

Farrell, H. 2012. The Consequences of the Internet for Politics. *Annual Review of Political Science* 15 (1): 35–52.

Gardbaum, S. 2008. Human Rights and International Constitutionalism. In *Ruling the World? Constitutionalism, International Law and Global Government*, ed. J. Dunoff and J. Trachtman, 233–257. Cambridge: Cambridge University Press.

George, E. 2019. *Digitalization of Society and Socio-political Issues*. London: Wiley.

Gill, Lex, Dennis Redeker, and Urs Gasser. 2015, November 9. *Towards Digital Constitutionalism? Mapping Attempts to Craft an Internet Bill of Rights*. Berkman Center Research Publication No. 2015-15. https://ssrn.com/abstract=2687120 or https://doi.org/10.2139/ssrn.2687120

Gillespie, Tarleton. 2018. *Custodian of the Internet*. London: Yale University Press.

Global Forum for Media Development. 2019. GFMD Statement on the Christchurch Call and Countering Violent Extremism Online. https://drive.google.com/file/d/1N4EwiM7eITD6plQrYqJI02NayvCiMCT-/vi

Global Network Initiative. 2008. Principles on Freedom of Expression and Privacy. https://www.globalnetworkinitiative.org/principles/index.php

Gorwa, R., R. Binns, and C. Katzenbach. 2020. Algorithmic Content Moderation: Technical and Political Challenges in the Automation of Platform Governance. *Big Data & Society* 7 (1).

Grey, T.C. 1977. Procedural Fairness and Substantive Rights. *Nomos* 18: 182–205.

Grimm, D. 2016. *Constitutionalism: Past, Present, and Future*. Oxford: Oxford University Press.

Habermas, J. 1992. *The Structural Transformation of the Public Sphere*. Polity Press.

Héritier, A., and D. Lehmkuhl. 2008. The Shadow of Hierarchy and New Modes of Governance. *Journal of Public Policy* 28 (1): 1–17.

Internet Rights and Principles Coalition. 2010. The Charter of Human Rights and Principles for the Internet. http://internetrightsandprinciples.org/site/charter/.

Jorgensen, Rikke Frank, and Lumi Zuleta. 2020. Private Governance of Freedom of Expression on Social Media Platforms: EU Content Regulation Through the Lens of Human Rights Standards. *NORDICOM Review: Nordic Research on Media and Communication* 41 (1): 51–68.

Kaefer, Florian, Juliet Roper, and Paresha Sinha. 2015. A Software-Assisted Qualitative Content Analysis of News Articles: Example and Reflections. *Forum Qualitative Sozialforschung / Forum: Qualitative Social Research* 16 (2).

Kaye, D., 2018. Promotion and protection of the right to freedom of opinion and expression. UN General Assembly, A/71/373 (6 September 2016), https://www.refworld.org/docid/57fb6b974.html

Kulesza, J. 2008. Freedom of Information in the Global Information Society: The Question of the Internet Bill of Rights. 1 University of Warmia and Mazury in Olsztyn Law Review 81.

Lessig, L. 2006. *Code 2.0*. New York: Basic Books.

Luhmann, N. 1975. *Potere e Complessità Sociale*. Milano: Il Saggiatore.

MacKinnon, Rebecca, Elonnai Hickok, Allon Bar, and Hai-in Lim. 2015. *Fostering Freedom Online: The Role of Internet Intermediaries*. UNESCO Internet Freedom Series.

Main, T.O. 2010. The Procedural Foundation of Substantive Law. *Washington University Law Review* 87: 801–847.

Moller, C. 2004. Transnational Governance Without a Public Law? In *Transnational Governance and Constitutionalism*, ed. C. Joerges, I.Y. Sand, and G. Teubner, 329–337. Oxford: Hart Publishing.

O'Connor, C., and J.O. Weatherall. 2019. *The Misinformation Age: How False Beliefs Spread*. Yale University Press.

Palladino, N. 2021a. The Role of Epistemic Communities in the "Constitutionalization" of Internet Governance: The Example of the European Commission High-Level Expert Group on Artificial Intelligence. *Telecommunications Policy* 45 (6): 1–15.

———. 2021b. Imbrigliare i giganti digitali nella rete del costituzionalismo ibrido. Spunti dall'approccio europeo alla governance dell'Intelligenza artificiale. *Comunicazionepuntodoc* 25: 123–140.

———. 2022. A 'Biased' emerging Governance Regime for Artificial Intelligence? How AI Ethics Get Skewed Moving from Principles to Practices. *Telecommunications Policy*. https://doi.org/10.1016/j.telpol.2022.102479.

Pasquale, F. 2015. *The Black Box Society: The Secret Algorithms that Control Money and Information*. Harvard University Press.

Peters, A. 2006. Compensatory Constitutionalism: The Function and Potential of Fundamental International Norms and Structures. *Leiden Journal of International Law* 19 (3): 579–610.

Reporters Sans Frontiers. 2018. International Declaration on Information and Democracy. https://rsf.org/en/global-communication-and-information-space-common-good-humankind

Santaniello, M., and N. Palladino. 2022. Discourse Coalitions in Internet Governance. In *Internet Diplomacy: Shaping the Global Politics of Cyberspace*, ed. A. Calderbrook and M. Marzouki. Rowman and Little fields.

Santaniello, M., E. De Blasio, N. Palladino, D. Selva, E. De Nictolis, and S. Perna. 2016. Mapping the Debate on Internet Constitution in the Networked Public Sphere. *Comunicazione politica* 17 (3): 327–354.

Santaniello, M., N. Palladino, M.C. Catone, and P. Diana. 2018. The Language of Digital Constitutionalism and the Role of National Parliaments. *International Communication Gazette 80* (4): 320–336.

Sciulli, D. 1992. *Theory of Societal Constitutionalism*. Cambridge: Cambridge University Press.

Smith, J. 2020. The Blackbox Problem in Content Moderation. *Journal of Online Platforms* 5 (2): 56–68.

Teubner, G. 2004. Societal Constitutionalism: Alternatives to State-Centred Constitutional Theory? In *Transnational Governance and Constitutionalism*, ed. C. Joerges, I.Y. Sand, and G. Teubner, 3–28. Oxford: Hart Publishing.

———. 2011. Transnational Fundamental Rights: Horizontal Effect? *Netherlands Journal of Legal Philosophy* 40 (3): 191–215.

———. 2012. *Constitutional Fragments: Societal Constitutionalism and Globalization*. Oxford: Oxford University Press.

Thornhill, C. 2011. *A Sociology of Constitutions: Constitutions and State Legitimacy in Historical-Sociological Perspective*. Cambridge: Cambridge University Press.

Waldron, Jeremy. 2020. The Rule of Law. In *The Stanford Encyclopedia of Philosophy*, ed. Edward N. Zalta, Summer 2020 ed. https://plato.stanford.edu/archives/sum2020/entries/rule-of-law/.

Walker, G.D.Q. 1988. *The Rule of Law: Foundation of Constitutional Democracy*. Melbourne University.

CHAPTER 5

Platform Policies Versus Human Rights Standards

Abstract This chapter empirically examines how five social media platforms—Facebook, Instagram, Twitter, TikTok and YouTube—deal with the content governance dilemma and the question of which human rights standard to apply when moderating user content. It builds on previous chapters' analyses of relevant human rights standards in international law and civil society-issued documents to elucidate to what extent substantial and procedural demands are met by the platforms. After an analysis of platform policies—specifically the human rights commitments included in them, the chapter examines substantive content moderation trends in a comparative way. Thereafter, procedural practices of content moderation including transparency reporting and automated content moderation are comparatively discussed. The chapter finds a relatively high degree of convergence among the platforms on a number of practices.

Keywords Platform policies • Human rights standards • Content moderation trends • Transparency reports • Automated content removal

© The Author(s) 2023
E. Celeste et al., *The Content Governance Dilemma*, Information Technology and Global Governance,
https://doi.org/10.1007/978-3-031-32924-1_5

93

5.1   Human Rights Commitments as a Window Dressing Strategy?

This chapter empirically examines the content moderation practices of four selected platform companies and five of their social media services—Facebook and Instagram (Meta Inc.), YouTube (Alphabet Inc.), Twitter (Twitter Inc.) and TikTok/Douyin (ByteDance Ltd.). This chapter is based on the analysis of norms in international law presented in Chap. 3 and the findings from the empirical analysis of civil society documents included in Chap. 4. It illustrates how social media platforms deal with the content governance dilemma outlined previously. In Chap. 3, we demonstrated to what extent the content governance of transnational platforms can be regulated or guided by international human rights law (to only a limited extent). The current chapter shows how four globally operating platform companies are dealing with this relative lack of strict guidance, but also with the nonetheless large amount of existing human rights writing and commentary. This chapter also illustrates that a more elaborate human rights standard developed by the international community and put into treaties by states in a multistakeholder process could be desirable, if only to address the gap between those companies that do more to protect human rights in their operations and those that do far less. The chapter also illustrates the value of civil society and multistakeholder charters and declarations that are at times directly cited to be an impetus towards a stronger human rights commitment of platforms. It also compares to what extent substantive and procedural demands raised in these documents are met by different platforms. The chapter only indirectly addresses the legislative codification of international human rights standards into national law and regional rules (e.g. the recently proposed EU Digital Services Act package). Instead, it takes the four platform companies and their services as the locus of the analysis. How these platforms *translate* general human rights commitments into platform policies and practices matters greatly for practical human rights protection online. Observing recent activities of these platforms closely allows us to tease out the different ways in which the content governance dilemma can be addressed, and it helps us to understand *why* platforms address it as they do.

When analysing how these platforms deal with content posted by users in connection to human rights norms, one must take into account both the formal commitments and statements of the platforms and the empirical—or sociological—reality of how platforms incorporate human rights standards into their processes—or not. Both potentially tell us something about the reasons for adopting, or failing to adopt, a strong human rights stance in content governance. Are human rights commitments a mere window dressing? How do the platforms structure their moderation processes and what moderation outcomes can be observed in relation to how civil society documents frame desirable moderation principles? To address these questions, this chapter is subdivided into three sections. In a first step, we explore the written platform policies on content moderation and show to what extent these documents include human rights language and an explicit commitment to human rights norms or a dedicated policy on human rights. In a second step, we explore how the *substantive* demands by civil society Internet Bills of Rights, focussing on what is seen to be legitimate exceptions from their central freedom of expression claim, are being realised through content moderation at the five platform services. Using the data published by the platforms themselves, we have focused on the human rights relevant moderation practices based on a framing by civil society documents, using the categories developed in Chap. 4. This helps us to connect a discussion about principles adopted in the content policies of platforms with the idea of an emerging convergence—or standard—for content governance. This convergence occurs both on the level of content moderation policy documents and on the level of substantive moderation outcomes. In a third step, we examine the *procedural* category of principles entailed in the Internet Bills of Rights. By looking at two specific principles and respective metrics, (1) the share of moderation decisions taken by automated systems such as AI technologies over time and (2) the increase of transparency reporting of platforms, we focus on some of the key principles demanded by civil society-issued documents. While both procedural principles tell us another story of converging to a standard across platforms, the impact on human rights is less clearly identifiable. The continuous struggle for greater human rights protection is as much needed as are suitable platform policies and moderation practices.

5.2   Platform Policies and Human Rights Commitments

That social media companies concern themselves with the *right* human rights standard for their content moderation operations is a relatively new phenomenon, much like the idea that platforms closely watch what users publish at all. For a long time, social media platforms happily took on the cloak of 'content intermediaries', which promote free speech and present little in terms of rulebooks to their users. The platforms in fact benefitted from regulations such as Section 230 of the US Communications Decency Act (CDA) and the E-Commerce Directive 2001 in the European Union (Citron and Wittes 2017; Kuczerawy and Ausloos 2015). These regulations allowed them to evade direct liability for content posted by users but obliged them to act when being notified of potential violations and infringing content. This allowed the companies to claim for their social media services the status of neutral tech companies that support (American) First Amendment protections by maximising free speech (and reach) on the Internet. This, almost libertarian, approach mirrors the early Internet ideals most eloquently captured in the Declaration of Independence of Cyberspace (Barlow 1996). This being said, when the four platforms started their operations, there were content limitations such as restrictions on pornographic content, copyrighted materials and spam. Only later would these be coded into their initial content policies.[1] Twitter, in 2009, in its very first iteration of the "Twitter Rules", still stated that

> each user is responsible for the content he or she provides [and thus], we do not actively monitor user's content and will not censor user content, except in limited circumstances described below. (Twitter 2009)

In the same document, Twitter provided a narrow set of exceptions to the focus on freedom of expression, most notably with regard to cases, indeed, referring to spam, pornography, privacy and copyright infringements. Twitter's early platform policies were an important step towards spelling out the rules for speech on the platform but they were less of a concern for policymakers or human rights groups as they are today.

Up until the early 2010s, the scope of platform content policies was relatively limited. Over time, as massive growth of the user base lifted the profile of social media platforms, they became more entangled in political affairs—shaping electoral politics, but also being affected by increasing demand and regulation (Barrett and Kreiss 2019). At the same time, these platforms also represented a viable source of personal data used by national

---

[1] For an overview of the early platform content policies, see the Platform Governance Archive, https://www.platformgovernancearchive.org/.

intelligence agencies, which also gave rise to a new form of surveillance capitalism (Zuboff 2019). Starting in the mid-2010s, a number of high-profile scandals further put social media platforms into the focus of policy-makers in the political capitals of the world. After the 2013 Snowden revelations about spy agencies and their ready access to data from social media companies, 2016 represents another inflexion point, with wide-spread discussions of misinformation on platforms following the US elections, while the 2018 Cambridge Analytica scandal brought further concerns about privacy and corrupted electoral processes in connection with data collected through Facebook (Hemphill 2019). The "techlash" (Hemphill 2019) that followed the scandals led to greater pressure to design more complex content policies and to innovate with regard to content moderation procedures. Amid this "turn to responsibility" (Katzenbach 2021), platforms moved further away from the notion of platform neutrality in matters of content, which represented a core ingredient of the rise of social media platforms. Instead, today, there exists a "broad consensus that platforms have responsibility for the content and communication dynamics on their services" (Katzenbach 2021, 3). This turn can also be detected through changes in public platform content policies. Twitter, for example, after a major revision of its Rules in June 2019, stated that its

> purpose is to serve the public conversation. Violence, harassment and other similar types of behaviour discourage people from expressing themselves, and ultimately diminish the value of global public conversation. Our rules are to ensure all people can participate in the public conversation freely and safely. (Twitter 2020)

This statement marks a dramatic shift from the initial free-speech absolutism of the platform's early days—and perhaps the days that lie ahead, after the acquisition of the platform by billionaire, Elon Musk.

Twitter is not an exceptional case in this matter. Other platforms have also developed substantial, and elaborate rulesets concerning the kind of content that can be posted on their sites. These platform policies are usually documented on public pages for users to consult. At times, such as in the case of Facebook and Instagram, these rulesets are flanked by transparency centres in the form of websites providing information on enforcement practices. These webpages are ostensibly geared to be of use to policymakers, journalists, members of organised civil society and academic researchers. This is appropriate because the written policies of large social media companies and their enforcement practices represent a comprehensive and powerful mode of governing communication on the Internet. That online communication is governed in such a way by private actors rather than public entities may seem "lawless" in its current state due to a perceived lack of legitimacy of platforms to rule (Suzor 2019), and it might amount to normative platform authoritarianism as argued in Chap. 2. Notwithstanding the unease many observers perceive, the facticity of "platform law" (Bygrave 2015; Celeste 2022; United Nations 2019) remains, and with it the dominant role intermediaries play in governing the Internet (Suzor 2019). Platform law is at play even where it is not published, where rules are kept secret or otherwise unavailable. There may be a number of reasons why rules are not public, including differential treatment of specific groups (as in the recently revealed separate content moderation for celebrities' content on Meta Inc.'s platforms),[2] the lack of codification (in the case of early platforms) or because content moderation is intertwined with state censorship (as, for instance, in the case of Chinese social media platforms).

In this context, it is important to highlight that an increased number of rules for content posted on these services does not necessarily amount to effective human rights protection. Instead, the growth of the number of rules *per se* can also stifle important values and rights such as equality or freedom of expression, which is so central to civil society Internet Bills of Rights. In an environment in which platforms are increasingly pushed to over-moderate to save themselves from legal peril or a public relations disaster, a multitude of rules that allow for speech to be removed may be an outright risk to freedom of speech. Ideally, at least in the framework of this book, platform rules directly refer to the human rights document, whose implementation they ought to support. However, when examining the four platform provider's content moderation rules, such immediate references cannot be found. This may well be due to the difficulty to simply copy and paste the content of international human rights documents, as discussed in Chap. 3. However, human rights are usually referred to by the platform services in some way. For our purposes and the remainder of

---

[2] A 2021 leak, the so-called Facebook Files, entailed information about Meta's XCheck program, which in late 2020 shielded at least 5.8 million important and celebrity users from the content moderation procedures applied to other users (see Horwitz 2021).

this section, we are most interested in *how* human rights are included in the platform policies of each of the four platform companies.

5.2.1   Meta

Meta Inc. is the parent company of two major social media services—Facebook and Instagram, which share a common set of policy documents. The popularity of the platforms—just under 3 billion people use Facebook every month and just under 1.5 billion use Instagram—is the foundation for the company's place among the largest companies globally by valuation (Statista 2022). WhatsApp, a messenger service, is another popular service owned by Meta, which is however not examined in this chapter. The content posted on Meta's platforms is governed by its "Community Standards", which were first published in 2007 (Facebook 2007). The document has greatly expanded from a little over 700 words in its first iteration to more than 19,700 words spread across several sub-pages as of late 2022, now including many explanations and examples (Meta 2022a). The Community Standards also apply to content posted on Instagram. Due to these services' enormous number of users, the rules may very well be one of the most effective tools to affect the enjoyment of human rights worldwide, true constitutional instruments of these online spaces (Celeste 2019). Hence, the pressure on the company by civil society activists and by the former UN Special Rapporteur on the Promotion and Protection of the Right to Freedom of Opinion and Expression, David Kaye, to adopt *a* human rights standard for its content moderation has been immense (Helfer and Land 2022; United Nations 2018, 2019). Starting in 2018 and 2019, Meta (then still known as Facebook Inc.) started to adopt human rights references in communications about content moderation by top management (Allan 2018; Zuckerberg 2019). This talk was then followed up with the creation of the Meta Oversight Board, which has its own charter and by-laws that explicitly put the board on a path to negotiate between the platform's Community Standards on the one hand and, external, human rights standards on the other hand, as we will discuss below. Legally speaking, it still holds true what the company communicated in 2018, that is, "we're not bound by international human rights laws that countries have signed on to" (Allan 2018). Nonetheless, in 2021, Meta gave itself its own corporate human rights policy, stressing both a commitment to the non-binding Ruggie Principles and other international law instruments:

> We are committed to respecting human rights as set out in the United Nations Guiding Principles on Business and Human Rights (UNGPs). This commitment encompasses internationally recognized human rights as defined by the International Bill of Human Rights—which consists of the Universal Declaration of Human Rights; the International Covenant on Civil and Political Rights; and the International Covenant on Economic, Social and Cultural Rights—as well as the International Labour Organization Declaration on Fundamental Principles and Rights at Work. (Meta 2022c)

As discussed in Chap. 3, the UNGPs are international soft-law standards that systematically address businesses, albeit indirectly. For a platform company like Meta to commit specifically to the Ruggie Principles should be a minimum standard, or as the former UN Special Rapporteur on the Promotion and Protection of the Right to Freedom of Opinion and Expression recommended:

> The Guiding Principles on Business and Human Rights, along with industry-specific guidelines developed by civil society, intergovernmental bodies, the Global Network Initiative and others, provide baseline approaches that all Internet companies should adopt. (United Nations 2018)

To what extent these listed commitments to international human and labour rights standards result in a coherent human rights standard, particularly considering the strong competing role of Meta's other values, is discussed further below.

5.2.2   Twitter

Twitter Inc.'s platform may be the go-to place for politicians, journalists, academics and others to communicate political messages, advertise their own latest products or publications and engage in shoulder-rubbing by other means. However, importantly, many more users engage in everyday conversations about their life, leisure and politics. Twitter has also developed into a tool for human rights defenders to speak to members of the media to create awareness for human rights abuses by governments and companies, allowing for messages unfiltered by the press, governments or non-governmental organisations. For instance, in 2011, in Cairo's Tahrir Square, activists and ordinary citizens connected to one another and to a global audience through their 'tweets', at the very least *supporting* the

2011 Egyptian Revolution. In early 2022, Twitter counted more than 400 million active monthly users globally (Statista 2022), and not all of them were bots (Milmo 2022). Twitter had been the favourite outlet of thoughts and supposed policy formulations by former US President Donald Trump. Like Facebook, Twitter faced a decision on how to continue the relationship with the (then-sitting) president after the storming of the US Capitol on January 6, 2021. Twitter decided to permanently suspend Trump's account (Guo 2021). Generally, what Twitter users can and cannot post online is regulated by Twitter Rules and additional documents. A dedicated human rights policy was not publicly available as of September 2022. Twitter claims that its commitment to user rights is based on a commitment to both the US Constitution and the European Convention on Human Rights (Twitter 2022a). It also refers to the fact that its content moderation is "informed (…) by works such as United Nations Principles on Business and Human Rights" (ibid.). However, unlike Meta, the company does not specify particular rights or further international human rights documents that could amount to a binding policy or standard.

5.2.3   TikTok

TikTok has been *the* recent quick starter among social media platforms globally. The platform's focus on short video clips as a core format harnesses the increased access to fast mobile data connections and suitable mobile phones to record scenes. The number of monthly active users more than doubled between the second quarter of 2020 and the second quarter of 2022, to an estimated 1.46 billion, according to the trade website Business of Apps (Iqbal 2022). In China, the mobile app and social media service is known as Douyin. While there had been pressure by the US government to spin-off its operations in the United States to a local joint venture, this did not directly occur. An attempted ban of TikTok by the previous administration was revoked by US President Biden (Kelly 2021). Being pressured to increase the protection of US data, TikTok's mother company ByteDance Ltd. arranged a deal with Oracle to store American user data within the country exclusively (The Guardian 2022). As of September 2022, TikTok's content policy, the "Community Guidelines" do not make any reference to human rights or international legal norms (TikTok 2022b). Nonetheless, in its transparency centre, after pointing to the fact that "as a global entertainment platform, TikTok spans most major markets except China, where ByteDance offers a different short-form video app called Douyin", the company published the following human rights statement:

> Technology is an essential gateway to the exercise of human rights. (…) Responsibility for upholding human rights is shared: while governments have the responsibility to protect human rights, TikTok and other businesses have a responsibility to respect those human rights. Respecting human rights is essential for TikTok to build and sustain trust among our employees, creators, advertisers, and others who engage with our company. Our philosophy is informed by the International Bill of Human Rights (which includes the Universal Declaration of Human Rights and the International Labour Organisation's Declaration on Fundamental Principles and Rights at Work) and the United Nations Guiding Principles on Business and Human Rights. As part of our commitment, we will strive to respect human rights throughout our business and will comply with applicable laws and regulations intended to promote human rights where we conduct business globally. We will continuously evaluate our operations to identify, assess, and address salient human rights risks; engage key stakeholders; and prioritise key areas where we have the greatest opportunity to have a positive impact. (TikTok 2022a)

The statement generally follows the Ruggie Principles but with a caveat that the platform would prioritise actions avoiding human rights based on where such actions would have the greatest benefit, rather than striving for an overall protection of human rights. Such a utilitarian approach to the balancing of rights and other objectives is interesting, particularly with regard to the processes and actors involved in such decisions. Based on the policy itself, little is to be expected in terms of aligning with the substantial and procedural demands by civil society voiced in various Internet Bills of Rights examined in Chap. 4.

5.2.4   YouTube

YouTube is a global video platform owned by Google Inc., which itself is a subsidiary of Alphabet Inc. Google ran with the slogan "don't be evil", which could also be found in the company's Code of Conduct—at least until it was removed from there in 2018 (Conger 2018). YouTube counts more than 2.5 billion monthly active users globally, which places it within the top four social media services, together with three services offered by

Meta Inc. (Statista 2022). Like the other platforms featured in this book, YouTube's very existence—particularly in countries with restricted public discourse—can be seen as a contribution towards enhancing freedom of expression and the right to information. Thus, when the platform was blocked by governments, human rights courts have repeatedly found that this blocking amounted to a human rights violation of the platform's users (Deutsche Welle 2015). However, YouTube has also been subject to allegations that it does not do enough to fight human rights violations (AccessNow 2020). Interestingly, as of September 2022, no explicit commitment to human rights can be found in YouTube's "Community Guidelines", which govern the kind of content that can be posted on the platform (YouTube 2022a). However, Google Inc. has a human rights policy that also applies to YouTube. Specifically, Google's policy asserts that the company finds orientation in internationally recognised human rights standards "in everything it does", adding a commitment to respect the rights included in the Universal Declaration of Human Rights and related treaties (Google 2022). The statement also specifically mentions the Ruggie Principles and, interestingly, the principles of the Global Network Initiative (GNI). The human rights policy also includes information on how Google and YouTube aim to implement these commitments, and thus translate the principles into moderation practices.

When observing how the four platforms discuss human rights in their policy documents, some stark similarities but also differences can already be made out. All five platforms (or their parent companies) include references to human rights, including specifically to the Ruggie Principles. Thus, on paper, one might say that a strong convergence on committing platforms to human rights standards has developed in the field, even if the scope of applicable human rights documents differs and differences in approach can be made out (e.g. TikTok's decidedly utilitarian approach). However, importantly, these human rights policies are distinct from content governance policies. The former are likely not integrated into the latter, in part due to the challenges that are posed by the application of any *one* human rights standard to content governance. As argued above, digital constitutionalism and, specifically, civil society Internet Bills of Rights are a potential catalyst to solving the content governance dilemma. Civil society advocates show platforms the way by balancing, in their documents at least, various human rights and good governance principles against each other. Consequently, the findings from Chap. 4, particularly if taken in their aggregate, can inform efforts to apply human rights standards to platform content moderation. Consequently, the next two sections investigate how the content governance practices of the featured platforms perform against the background of these civil society demands. The next section focuses on substantial demands by civil society and the quantitative outcomes of platform content moderation.

5.3   SUBSTANCE MATTERS! PLATFORM MODERATION OUTCOMES VERSUS CIVIL SOCIETY DEMANDS

There are many who rightly emphasise the importance of process when it comes to evaluating the content governance of social media platforms (Kettemann and Schulz 2020; Klonick 2018; Suzor et al. 2018). However, as we show below, there is value in gauging to what extent comparative substantive enforcement outcomes relate to the civil society demands, if we assume that these demands, particularly in aggregate, are an important interpretation of how a human right-based platform governance regime should be designed. Addressing the three categories of "prevention of harm", "protection of social groups" and "public interest" as outlined in Chap. 4, this section considers which of these categories is most often used to justify limitations on the chief principle of freedom of expression. First, we categorised the substantive principles by which the four platforms organise content moderation into the three derived categories. The principles represent only the reported substantive principles for which data is available for analysis. Copyright infringements and related moderation principles are excluded from the analysis, due to differential reporting of data by platforms.[3] Table 5.1 shows which substantive principles can be found within the three mentioned categories, for all four platform companies, the information is limited to their reported data for 2021 (Meta 2022b; TikTok 2021a; TikTok 2021b; TikTok 2022c; TikTok 2022d; Twitter 2022b; YouTube 2022b).

Following the categorisation from Chap. 4, this chapter aims to show where platforms' respective focus lies. This framework is here first applied to Meta's two platforms and data for 2021, the last complete year for which data is available. In the following, the same framework is applied to Twitter, TikTok and YouTube.

---

[3] A recent report offers insights into substantive data on copyright moderation and copyright actions reporting by major platforms over time (Quintais et al. 2022).

Table 5.1 Substantive content moderation principles and categories from civil society documents

Categories	Principles reported on in transparency reports (2021)
Meta	
Prevention of harm	Bullying and harassment, suicide and self-injury, dangerous organisations: organised hate, dangerous organisations: terrorism, violence and incitement
Protection of social groups	Child endangerment: nudity and physical abuse, child endangerment: sexual exploitation, child nudity and sexual exploitation, adult nudity and sexual activity, violent and graphic content, hate speech
Public interest	Regulated goods: firearms, regulated goods: drugs
Twitter	
Prevention of harm	Abuse/harassment, hacked materials, impersonation, non-consensual nudity, private information, promoting suicide or self-harm, terrorism/violent extremism, violence
Protection of social groups	Child sexual exploitation, hateful conduct, sensitive media
Public interest	Civic integrity, COVID-19 misleading information, illegal or certain regulated goods or services, manipulated media
TikTok	
Prevention of harm	Harassment and bullying, suicide, self-harm and dangerous acts, violent and graphic content, violent extremism
Protection of social groups	Adult nudity and sexual activities, hateful behaviour, minor safety
Public interest	Illegal activities and regulated goods, integrity and authenticity
YouTube	
Prevention of harm	Promotion of violence and violent extremism, harmful or dangerous content, harassment and cyberbullying
Protection of social groups	Child safety, nudity or sexual content, violent or graphic content, hateful or abusive content
Public interest	Spam, misleading content, scams

Meta's Community Standards apply to the platforms Facebook and Instagram. Their content policy is informed by the organisation's self-proclaimed "core values", which emphasise their aim to create "a place for expression and giving people voice" (Meta 2022a). What limits free expression are four values—authenticity, safety, privacy and dignity—and the application of copyright rules and national law. The focus on freedom of expression is absolutely consistent with the emphasis on this principle by civil society, as we observed in Chap. 3. The Community Standards are structured into six chapters, of which the first four outline restrictions on content based on Facebook's core values; the fifth chapter affirms intellectual property and its protection, whereas the sixth chapter outlines which user requests Meta complies with, including those to protect children and youth. The first four chapters currently entail a total of 21 principles defining content that must not be posted on the platform (Meta 2022a). Each principle is described in some detail, some with bullet-pointed lists of what constitutes an offence to be removed.

For the year 2021, Meta issued content moderation transparency reports covering 15 categories of content that can cause an action to delete content (apart from copyright-related actions and actions based on national legal requirements). These categories do not neatly fit the 21 principles of the Community Standards. For instance, the principle not to share "Restricted Goods and Services" includes goods such as weapons, drugs, blood, endangered animals, weight loss products, historical artefacts or hazardous goods and materials. Reporting, however, is only done for drugs and weapons. Two reporting categories—prohibitions on fake accounts and on spam—are arguably not as closely associated with the three most important categories of civil society demands; they will not be discussed in this analysis. In addition, reporting on these is only available for Facebook, lowering the number of reporting categories for Instagram to 13. Data for 2021 is available separately for Facebook and Instagram (downloaded through Meta's Transparency Center, see Meta 2022b). Furthermore, data for "Child nudity and sexual exploitation" is only reported for the first quarter of 2021. Starting from the following quarter, Instagram and Facebook's data differentiate between "Child endangerment: Nudity and physical abuse" and "Child endangerment: Sexual exploitation" when reporting moderation actions in this area. Table 5.2 shows each category from the civil society documents and the corresponding principles from the Community Standards on which transparency reporting occurs. For some quarters, no data is reported for one or the other platform.

Differences between Facebook and Instagram in terms of relative share among the justifications to limit the core value of 'voice' are distinctive. Instagram users have been moderated more commonly based on justifications of preventing bullying, suicide and self-injury, adult nudity and the advertisement of drugs. On Facebook, there are relative shares of moderation due to child sexual exploitation, in relation to terrorist organisations and hate groups, and due to gun offerings.

Table 5.2 Moderation outcomes and civil society categories, Facebook and Instagram (2021)

Category/principle	Content actions (FB)	Content actions (IG)	Share of total (FB)[a]	Share of total (IG)
Prevention of harm	**150,700,000**	**51,086,900**	**25.74%**	**31.48%**
Bullying and harassment	34,100,000	24,500,000	5.83%	15.10%
Suicide and self-injury	36,600,000	17,000,000	6.25%	10.48%
Dangerous organisations: organised hate	19,600,000	1,330,100	3.35%	0.82%
Dangerous organisations: terrorism	34,400,000	2,356,800	5.88%	1.45%
Violence and incitement	26,000,000	5,900,000	4.44%	3.64%
Protection of social groups	**416,500,000**	**105,580,600**	**71.15%**	**65.06%**
Child endangerment: nudity and physical abuse	5,900,000	1,968,200	1.01%	1.21%
Child endangerment: sexual exploitation	66,600,000	5,600,000	11.38%	3.45%
Child nudity and sexual exploitation	5,000,000	812,400	0.85%	0.50%
Adult nudity and sexual activity	126,700,000	42,000,000	21.64%	25.88%
Violent and graphic content	115,900,000	29,300,000	19.80%	18.05%
Hate speech	96,400,000	25,900,000	16.47%	15.96%
Public interest	**18,200,000**	**5,616,100**	**3.11%**	**3.46%**
Regulated goods: firearms	6,000,000	516,100	1.02%	0.32%
Regulated goods: drugs	12,200,000	5,100,000	2.08%	3.14%
Not categorised	**10,190,500,000**	**N/A**	–	**N/A**
Fake accounts	6,500,000,000	N/A	–	N/A
Spam	3,690,500,000	N/A	–	N/A

[a]Excluding the moderation categories of fake accounts and spam, for which there is no data from Instagram and which dwarf the remainder of the categories in the case of Facebook (94.6% of total content actions of the 15 reported on categories)

Table 5.3 shows data for Twitter for the entire year of 2021 (Twitter 2022b). The platform's transparency reporting differentiates content policy-related sanctions into 'content deletions' and 'account suspensions', both adding up to 'account actions'. Across different categories derived from the Internet Bills of Rights, account suspension shares differ. 'Public interest'-related enforcements and account suspensions make up

Table 5.3 Moderation outcomes and civil society categories, Twitter (2021)

Category/principle	Account actions	Account suspensions	Share of total (actions)	Share of total (suspensions)
Prevention of harm	**3,338,114**	**670,047**	**34.13%**	**26.19%**
Abuse/harassment	1,984,204	182,536	20.29%	7.13%
Hacked materials	143	0	0.00%	0.00%
Impersonation	398,490	368,625	4.07%	14.41%
Non-consensual nudity	58,471	15,660	0.60%	0.61%
Private information	64,895	5741	0.66%	0.22%
Promoting suicide or self-harm	753,243	18,818	7.70%	0.74%
Terrorism/violent extremism	78,668	78,667	0.80%	3.07%
Protection of social groups	**5,990,781**	**1,679,348**	**61.25%**	**65.64%**
Child sexual exploitation	1,055,669	1,050,751	10.79%	41.07%
Hateful conduct	2,010,891	238,150	20.56%	9.31%
Sensitive media	2,773,618	282,616	28.36%	11.05%
Violence	150,603	107,831	1.54%	4.21%
Public interest	**452,747**	**209,058**	**4.63%**	**8.17%**
Civic integrity	674	27	0.01%	0.00%
COVID-19 misleading information	51,947	1993	0.53%	0.08%
Illegal or certain regulated goods or services	399,983	207,038	4.09%	8.09%
Manipulated media	25	0	0.00%	0.00%

46% of actions; the rate is significantly lower for the 'prevention of harm' (20%) and 'protection of social groups' (28%) categories. The high 'public interest' share is due to a relatively more hard-line approach to moderation of displays and offers of 'regulated goods or services', as part of which accounts are more often suspended. Notably, almost all violations of the principles against 'terrorism/violent extremism' and 'child sexual exploitation' led to account suspensions rather than mere content removal.

The data for Twitter shows that more than 60% of account actions occurred to "protect social groups". Within that category "child sexual exploitation" makes up the by far largest reason for moderation actions.

Table 5.4 shows data for TikTok for all four quarters of 2021. Data availability is for "video removals" in these time intervals, rather than "content actions" (Meta) or "account actions" and "account suspension"

Table 5.4 Moderation outcomes and civil society categories, TikTok (2021)

Category/principle	Share video removals			
	Q1	Q2	Q3	Q4
Prevention of harm	**14.20%**	**13.10%**	**11.90%**	**13.90%**
Harassment and bullying	8.00%	6.80%	5.30%	5.70%
Suicide, self-harm and dangerous acts	5.70%	5.30%	5.70%	7.40%
Violent extremism	0.50%	1.00%	0.90%	0.80%
Protection of social groups	**62.70%**	**65.20%**	**70.90%**	**66.00%**
Adult nudity and sexual activities	15.60%	14.00%	11.10%	10.90%
Hateful behaviour	2.30%	2.20%	1.50%	1.50%
Minor safety	36.80%	41.30%	51.00%	45.10%
Violent and graphic content	8.00%	7.70%	7.40%	8.50%
Public interest	**23.1%**	**21.70%**	**17.10%**	**21.10%**
Illegal activities and regulated goods	21.1%	20.90%	16.60%	19.50%
Integrity and authenticity	2.00%	0.80%	0.50%	0.60%

(Twitter). Differences of metrics reported here are based on differences in platform reporting. In addition, TikTok only provides quarterly figures in their Transparency Center. In general, the degree of detail is relatively low for TikTok. However, the platform does offer data on content actions by countries, at least for a small number of countries, potentially useful information only few other platforms (such as YouTube) report on in detail. TikTok's video removal data relates to only nine moderation principles for posted content. Not included in the table is more detailed information about spam and fake accounts and engagement, not reported in this detail by other platforms. For instance, in the last quarter of 2021, TikTok "prevented" more than 152 million spam accounts, removed more than 46 million spam videos as well as 442 million fake followers, 11.9 billion fake likes and more than 2.7 billion fake follow requests (TikTok 2022d).

The data for TikTok shows that, like the previous three platform services, protection of social groups makes up the largest share of the three categories derived from civil society demands. The included limitations on freedom of speech are mostly justified with "minor safety"; here this single moderation principle amounts to more than half of all deleted videos (at least during the period July and September 2021). It appears that this justification is often used, perhaps due to the current character of TikTok used by younger users.

YouTube's Community Guidelines entail 21 moderation principles or sub-guidelines on its website (YouTube 2022a). One of these, however, is itself a list of other guidelines pertaining to four reasons for moderation that appear on the face of it to be too small to be an entire principle alongside the others. Interestingly, the guideline category of misinformation entails three sub-guidelines (or principles) prohibiting general misinformation, misinformation related to elections and medical misinformation related to the COVID-19 pandemic. This, perhaps once again, illustrates the effect world events have on the policies themselves, if not also their enforcement. YouTube's report on "YouTube Community Guidelines enforcement" entails only data on eight principles (YouTube 2022b). Table 5.5 shows how the substantive moderation decisions the platform reports on for 2021. Like the other platforms examined here, YouTube's reporting shows a strong—or even stronger—quantitative emphasis on removing videos that may (be used to) hurt (the sensibilities of) certain groups, including children and protected groups. Spam video removal is included in the data presentation here, because of its bundling up with other moderation principles such as misleading content and scams.

There is a lack of comparative research into substantive content moderation outcomes of Platforms. We ventured to conduct a comparison utilising a broad framework developed from civil society demands for one year of reported data. These demands, we argue, can help platforms understand what rights and principles should be considered when limiting

Table 5.5 Moderation outcomes and civil society categories, YouTube (2021)

Category/principles	Share video removals			
	Q1	Q2	Q3	Q4
Prevention of harm	**4.74%**	**15.40%**	**13.33%**	**18.62%**
Promotion of violence and violent extremism	0.91%	6.9%	4.07%	1.90%
Harmful or dangerous content	2.22%	4.80%	4.58%	8.11%
Harassment and cyberbullying	1.61%	3.70%	4.68%	8.61%
Protection of social groups	**87.30%**	**70.40%**	**76.70%**	**72.27%**
Child safety	54.03%	29.90%	32.45%	31.53%
Nudity or sexual content	16.63%	22.40%	18.72%	18.42%
Violent or graphic content	15.73%	16.80%	23.70%	19.92%
Hateful or abusive content	0.91%	1.40%	1.83%	2.40%
Public interest	**7.96%**	**14.10%**	**9.97%**	**9.11%**
Spam, misleading content, scams	7.96%	14.10%	9.97%	9.11%

speech on their platforms. The clear downsides of such an analysis are, first, the differences in how many moderation principles are actually reported on—relative to the number of principles entailed in the platforms' content policies, and, second, the differences of what metrics are reported on. Regarding the latter, we see removals of videos (TikTok and YouTube), "content actions" (Facebook, Instagram, Twitter), and "account actions" (Twitter) as the dominant metrics in this space. Some platforms report on additional metrics whose use would, however, have made comparison even less viable.

With these caveats in mind, we find that, surprisingly, the shares between the three categories and for all five platform services are relatively similar. Table 5.6 shows that around two-thirds of all reported (non-spam, non-fake account) moderation actions are associated with the protection of social groups (range: 61.25% to 72.27%). Between 13.90% and 34.13% of moderation actions occurred to prevent harm, while between 4.63% and 21.10% of reported moderation decisions are categorised to be in the public interest. The deviations between platform services are certainly less than we would have expected. This is likely the case for two possible reasons. First, users' behaviour could be assumed to be relatively similar across platforms. This would mean that social media users globally conduct themselves on social media platforms in such a way as to require moderation in similar ways, say on TikTok as on Instagram. Some users, independently of which platform they are on, harm each other and post videos and other media that are deemed to be inappropriate for certain viewers, or they engage in behaviour that is regulated such as the sale of drugs. The other possible reason or the similarity observed has more to do with the reactions of the platforms to user behaviour. Political and market forces, including the recent techlash, have apparently impacted the content of the platform policies and the moderation processes in such a way that substantial moderation foci are relatively similar across platforms.

Table 5.6 Overall share of reported moderation actions by category, all five platform services (2021)

Category	Facebook	Instagram	Twitter	TikTok (Q4)	YouTube (Q4)
Prevention of harm	25.74%	31.48%	34.13%	13.90%	18.62%
Protection of social groups	71.15%	65.06%	61.25%	66.00%	72.27%
Public interest	3.11%	3.46%	4.63%	21.10%	9.11%

We suggest that the degree of similarity is indeed explained not just by coincidence. Such a suggestion arguably requires the assumption of relatively similar behaviour of users across platforms. It should also be noted that, where geographic differences in usership exist, these might have a slight impact on the overall trends. These effects notwithstanding, we argue that platforms converge in their global content moderation around a standard affected by public pressure.

To illustrate convergence, which describes an ongoing process, the moderation principle relating to the promotion of violence can be explored in some detail. Even before 2021, platforms usually had policies in place that would outlaw incitement to engage in violence. However, paying respect to the right of freedom of expression, platforms have been relatively less strict in their enforcement. This changed dramatically in January 2021 and shortly afterwards. There is evidence that the January 6 US Capitol attack strongly affected platform policies and practices. On that day, violence erupted around the building that houses both chambers of the US parliament, right when parliamentarians were to certify the results of the November 2020 general election. Both Twitter and Meta banned accounts of the then-US President Trump, who was identified as inciting and condoning the violence by way of his posts during and in the aftermath of the attack on Capitol Hill in Washington. Other platforms were quick to react rhetorically, with YouTube announcing that "due to the disturbing events that transpired yesterday, and given that the election results have now been certified, starting today any channels posting new videos with false claims in violation of our policies will now receive a strike" (Ha 2021). The adaptation of content moderation principles took a bit longer then, often pushed by external actors. For instance, in May 2021 the ban on Trump's accounts was in principle confirmed by Meta's Oversight Board, which took the case and decided that the decision made by the company was to be upheld. However, the Oversight Board argued that it was "not appropriate for Facebook to impose the indeterminate and standardless penalty of indefinite suspension" (Oversight Board 2021). Thereafter, the platform's Community Standards were substantially revised. The new version of the content policy included thinly veiled references to the riot at the Capitol and the role Trump played in the incitement of violence. Specifically, the late January 2021 version of the Community Standards prohibits content that makes "implicit statements of intent or advocacy, calls to action, or aspirational or conditional statements to bring armaments to locations, including but not limited to places of worship, educational facilities, polling places, or locations used to count votes or administer an election (or encouraging others to do the same)" (Facebook 2021). Less elaborate changes occurred faster. The term "incitement" was added to the Twitter Rules in January 2021, now stating that "content that wishes, hopes, promotes, *incites*, or expresses a desire for death, serious and lasting bodily harm, or serious disease against an entire protected category and/or individuals who may be members of that category" (Twitter 2021, emphasis added). This change, however, mimics the language of the day without being as directly related to the January 6 Capitol attack.

By mid-2021, all platform services included in our analysis entailed some reference to the principle of "incitement to violence", as is attested by Table 5.7. For Facebook and Instagram, data on the new principle is only available for the second half of the year 2021, quickly making up between 5 and 10% of overall moderation actions on the two platform services. The data further shows how the prohibition on incitement of violence was relatively more often invoked as a reason to take a content moderation action for some of the platforms studied here. This increase occurred on a low level in the case of TikTok and, slowly but strongly, in the case of YouTube, with a fall-off in terms of relative share of moderation actions in the fourth quarter of the year. In the case of Twitter, there was no significant change between the first and second half of 2021.

Changes in policies and moderation outcomes for one specific principle illustrate how a further convergence towards a *common standard* adopted by platforms can occur. These changes transpire due to public and political pressure to secure certain human rights—here, the right to life and the right to democratic elections. Civil society documents, including the GNI Principles and the Santa Clara Principles, have an impact when the platforms grasp for solutions to their policy and enforcement woes (as

Table 5.7 Share of reported moderation actions for incitement of violence, all five platform services (2021)

Platform	Principle	Q1	Q2	Q3	Q4
Facebook	Violence and incitement	N/A	N/A	8.81%	9.32%
Instagram	Violence and incitement	N/A	N/A	6.98%	5.93%
Twitter	Terrorism/violent extremism	0.86%		0.74%	
TikTok	Violent extremism	0.50%	1.00%	0.90%	0.80%
YouTube	Promotion of violence and violent extremism	0.91%	6.9%	4.07%	1.90%

indicated by them being cited in relation to human rights policies). However, not all global constituents have the same influence on shaping this standard; great tragedies could remain without an impact on policies if it was not for strong advocacy organisations to engage in reporting about platform failings. As examples from India and Myanmar show, platforms have been slow to adopt effective human rights-respecting policies and to conduct impact assessments (Al Ghussain 2022; Amnesty International 2022). Importantly, substantive moderation outcomes are not just affected by changes in policies. In fact, process matters quite a bit for the enjoyment of human rights, attested by the 40 civil society documents analysed previously. The next section examines two of these six core demands entailed in the Internet Bills of Rights in some more detail.

5.4   Process Matters! Platform Moderation Processes Versus Civil Society Demands

The substantive moderation outcomes discussed above tell us only as much about how civil society demands, which we understand as in their aggregate as a reasonable approach to how human rights-based content governance ought to be implemented, are actually met in practice. While the principles demanded, such as protection from hate and the protection of democratic elections create an important foundation for a human-rights-respecting moderation system, a suitable process is required to enable effectiveness, fair treatment and transparency in content moderation. Chapter 3 identified 32 procedural principles in 3 categories representing civil society demands with regard to the process of content moderation. In order to examine to what extent the empirical moderation practices of the four platform services adhere to demands by civil society, and to see whether they again converge, this section focuses on two of these principles: the *limitation of automated content moderation* and *transparency*. This focus is aimed to allow for more in-depth analysis.

5.4.1   Curbing Automated Content Moderation?

A relatively high number of civil society documents entail demands for limitations to automated content moderation. These demands are likely driven by the principled idea that every user and their posts should be evaluated by another thinking human being and not by a cold machine that perhaps does not really understand the point of the joke or the circumstance of the post in the first place. Examples of such false positives include the breast cancer awareness post in Brazil removed by automated systems for infringement on the company policy against nudity, even though it was clearly stated that the nude female breasts were shown for that exact, and permitted, purpose (Oversight Board 2020). In another case, a human moderator penalised a user for posting an Iranian protest slogan amid the 2022 protests against the Iranian government. The user, appealing to the initial decision by Meta, did not receive a decision by another human moderator, but an automated system closing the appeal (Oversight Board 2022). Civil society documents reviewed by us demonstrate that there is a hope that limitations on 'automated', 'proactive' or 'AI-based' moderation may help to reduce false positives, thereby strengthening freedom of expression.

All five platform services studied for this chapter rely on automation in their content governance systems. However, Twitter, at least until late 2022, did not report the number or the share of content removals triggered by automated detection of a policy violation. Demands of civil society concerning automated moderation, as seen in Chap. 3, are diverse. The Global Forum for Media Development's stance against any kind of automated content moderation, and the demands found in the "Charter of Digital Fundamental Rights of the European Union" drafted by members of the German civil society both amount to a demand for a right not to have decisions over humans be made by algorithmic systems. Such a demand is certainly difficult to reconcile with the strikingly pervasive use of automated moderation systems in content moderation. The results of the analysis are displayed in Table 5.8. The data shows that the four reporting services heavily rely on automated moderation to remove content from their platform. The striking exception is TikTok, which only removes about half of the videos it deems to violate its content policies upon a

Table 5.8 Share of automated moderation actions of total actions

Platform	Framing	Share of automation	Reference period
Facebook	Proactive detection	94.20%	Q3/2022
Instagram	Proactive detection	94.70%	Q3/2022
Twitter	N/A	N/A	N/A
TikTok	Videos removed by automation	48.02%	Q3/2022
YouTube	Automated flagging	94.52%	Q3/2022

prompt by an automated system. It can thus be concluded that automated systems take over a large share of the moderation workload, and they take (or at least took) on tasks even when a user appealed to a decision.

On the other hand, since human moderators would take longer to react to (automatically) flagged content, proactive moderation means that less material presumed to infringe platform policies will be viewed by users. Here, automated moderation and the call for it or a rejection become a balancing act between competing rights. Chiefly among others, the desire not to be over-moderated (as in the cases in Brazil and Iran) and to exercise freedom of expression. On the other hand, the 'right' to not see violent, hateful, sexual or privacy-infringing content and to be protected from online incitement of violence. Under-moderation, this is the tenor of the past several years, can have grave consequences for individuals and entire communities (Amnesty International 2022). Table 5.9 shows data from YouTube, illustrating the effects of automated moderation on the views of potentially policy-infringing content. Shown below is data for YouTube across a period of three years (2020–2022). Data is displayed for the third quarter of the year each and then annual intervals of data back to the earliest third quarter data available (YouTube 2022b).

The data suggests—not surprisingly—that automated detection decreases the share of videos removed for content policy violations ever seen by users. Videos picked up by other detection sources (for YouTube this means users, organisations or governments flagging content) are usually seen by people. In the case of many classes of content, such immediacy has great value. Abhorrent violence, pornography and terrorist propaganda may arguably not be suitable for young users. To wait for moderators to pick up the lead may well mean thousands or millions view content that will eventually be removed for policy violations. Still lacking explanation, the share of videos never seen by users decreased over time, from the third quarter of 2020 to the third quarter of 2022. In any case, much depends on the quality of the automated detection, which will likely

Table 5.9 Share of removed videos not viewed, for automated and other detection, YouTube (2020–2022)

Detection type	Q3/2020	Q3/2021	Q3/2022
Automated detection	45.2%	38.7%	38.3%
All other detection sources	2.7%	0.8%	3.6%

matter when balancing between over-moderation and under-moderation of platforms. This in turn depends on the quality of training data stemming from human moderators, which may be biased in a number of ways (Binns et al. 2017).

The over-time comparison of automation rates is an interesting indicator to observe trends and their stability of the algorithmic moderation. Arguably, such comparison can show how algorithmic moderation 'learns', taking over a larger share of the initial detection work from users and other actors. Table 5.10 shows data for the platform Facebook across a period of five years (2018–2022). Data is displayed for the third quarter of the year each and then annual intervals of data back to the earliest third quarter data available.

The data shows that the level of automated moderation is generally very high throughout the principles of Meta's Community Standards

Table 5.10 Share of proactive detection by category and year, Facebook (2018–2022)

Category/principles	Q3/2018	Q3/2019	Q3/2020	Q3/2021	Q3/2022
Prevention of harm					
Bullying and harassment	14.8%	16.2%	31.0%	59.4%	67.8%
Suicide and self-injury	N/A	96.8%	95.7%	99.0%	98.6%
Dangerous organisations: organised hate	N/A	N/A	97.8%	96.4%	94.3%
Dangerous organisations: terrorism	99.3%	98.5%	99.8%	97.9%	99.1%
Violence and incitement	N/A	N/A	N/A	96.7%	94.3%
Protection of social groups					
Child endangerment: nudity and physical abuse	N/A	N/A	N/A	97.1%	97.5%
Child endangerment: sexual exploitation	N/A	N/A	N/A	99.1%	99.5%
Child nudity and sexual exploitation	99.1%	99.5%	99.5%	N/A	N/A
Adult nudity and sexual activity	97.3%	98.8%	98.2%	98.8%	96.9%
Violent and graphic content	96.7%	99.0%	99.5%	99.4%	99.1%
Hate speech	52.9%	80.6%	94.8%	96.5%	90.2%
Public interest					
Regulated goods: firearms	N/A	97.6%	96.2%	96.7%	98.3%
Regulated goods: drugs	N/A	93.8%	91.7%	94.1%	94.8%
Not categorised					
Fake accounts	99.6%	99.6%	99.4%	99.8%	99.6%
Spam	99.7%	99.9%	99.9%	99.6%	98.5%

reported on. The overall rate of automation has increased over the last five years. For principles for which the automation rate has been relatively low still in 2018, such as principles against bullying and harassment, as well as hate speech, the automation shares have swiftly risen (from 14.8% to 67.8% for the former, and from 52.2% to 90.2% for the latter). These two principles illustrate how technically challenging the detection of hate speech, bullying and harassment are, given such expressions' contextual character. For other principles, the rate of automation has been consistently above 96–99% over the same period, amounting to the overall automated moderation of 94.2% referred to in Table 5.8. The data reported on a principle basis clearly shows that there is even a tendency away from the demand by civil society documents that automated content moderation should be limited to "manifestly illegal" content (and perhaps spam). In addition, as pointed out above, although demanded by some civil society documents, not all automated content decisions are being reviewed by a human.

Far from a limitation of automated moderation, the platforms studied here have extended their automated detection mechanisms and scaled them up. While data for Twitter is not readily available in their transparency report, it can be assumed that the service does not differ from the others on this indicator. We see once again that platforms become more similar and converge on the notion of near-complete automation of moderation, with moderators taking care of appeals (if at all). TikTok lags behind relatively speaking, but this might merely be a snapshot. The platform's overall automation rate has increased from 33.91% in the third quarter of 2021 to 48.02% in the third quarter of 2022 (TikTok 2022e). Whether this affords TikTok more appreciation by civil society is doubtful. The demands of civil society, as discussed in Chap. 3, are clearly not met. Neither is only "manifestly illegal" content being automatically detected, for instance, through a hash procedure as often done with copyrighted and terrorist content (Gorwa et al. 2020). Instead, increasingly, automation dominates moderation across content categories. A number of cases, in which Meta's Oversight Board has ordered the company to improve automated detection, shows that there are still regular and decisive failings of automated moderation even where, arguably, the most extensive set of training data should be available (Oversight Board 2020, 2022). As this subsection shows, being able to judge platforms on their self-reported data is key to understanding empirical developments and how they relate to any standard for content moderation extrapolated from civil society-authored Internet Bills of Rights. The following subsection shows how transparency reporting has also converged on a relatively extensive standard.

5.4.2   Transparency Reporting: Which Standard to Adopt?

Transparency is a core principle demanded in 16 of the civil society charters. For a platform to be transparent about its content moderation allows for others to scrutinise it, including but not limited to the question of whether the platform promotes and protects human rights. Why—apart from their human rights commitments, do platforms engage in activities that foster transparency and thus accountability? The goal of platforms when engaging in transparency-increasing measures—such as the creation of transparency reports and transparency microsites (transparency centres) that bring together various metrics and by engaging researchers and others—is to gain legitimacy. Transparency reports have become a key tool "to cultivate legitimacy with users and civil society organizations" (Suzor et al. 2018, 393). Legitimacy relates to the "right to govern" in the eyes of the users (the governed) but also, as a response or pre-emptive measure to public regulation, in the eyes of politically powerful stakeholders. Indeed, increasingly, regulators prescribe how platforms are required to report about their content moderation practices. India's Information Technology (Intermediary Guidelines and Digital Media Ethics Code) Rules of 2021 require larger platforms to produce monthly reports about complaints and actions taken (Tewari 2022). The Digital Services Act (DSA) and the Platform Accountability and Transparency Act (PATA) are respectively a recently adopted EU regulation and a US legislative proposal that would increase transparency requirements for platforms significantly. Transparency can further be enhanced by providing data on content moderation to academic researchers. Consequently, regulators increasingly perceive "access to data for empirical research (...) as a necessary step in ensuring transparency and accountability" (Nonnecke and Carlton 2022, 610). The DSA specifically "seeks a new level of granularity in transparency surrounding content moderation practices", surpassing previous national transparency reporting requirements such as the bi-annual requirement of the German NetzDG and India's transparency rules (Tewari 2022). Less in the focus of public attention yet already codified are transparency reporting standards for platforms towards their business partners as part of the EU's Platform-to-business Regulation of 2019

(European Union 2019). Based on this, Meta now regularly reports to their advertisers not only the number of complaints lodged against decisions and the type of complaint, but also the average time to process such appeals.

With regard to copyright-related notice-and-takedowns, additional voluntary transparency practices exist. For instance, the Lumen project at Harvard's Berkman Klein Center for Internet & Society collects and makes available DMCA takedown notices from those who receive them. This allows researchers and others to gain an understanding of individual practices and overarching trends. As of late 2021, Lumen included more than 18 million notices, most of them copyright-related, from companies such as Wikipedia, Google (including YouTube) and Twitter (Lumen 2022). Pending the passage of some of the more stringent legislative proposals, what is the level of transparency if platforms are being compared? Until 2019, the Electronic Frontier Foundation produced an annual report in which the content moderation practices of 16 online platforms were compared based on six overarching categories such as transparency about government takedown requests, transparency about content removal based on the platform's policies, transparency about appeals and even endorsement of one of the civil society-issued documents, the Santa Clara Principles (Crocker et al. 2019). In 2019, for the last iteration of the report, Reddit was able to receive a star in all six categories with Facebook, Instagram, Vimeo and Dailymotion performing particularly poorly.

Ranking Digital Rights produces an annual Big Tech Scorecard, which evaluates the corporate accountability of 14 (2022a) large digital platforms from the United States, China, South Korea and Russia, subdivided by offered services, such as Facebook and Instagram for Meta Inc. (Ranking Digital Tech 2022a). The report includes indicators on content moderation transparency reporting in its section on freedom of expression, such as the reporting of "data about government demands to restrict content or accounts", and data about platform policy enforcement. Overall, in that section, the report finds that Twitter "took the top spot, for its detailed content policies and public data about moderation of user-generated content" (Ranking Digital Tech 2022b). Table 5.11 shows an excerpt of the results for the subcategory of algorithmic transparency, also relevant for the proceeding subsection of this chapter.

The data in Table 5.11 suggests generally low scores in algorithmic transparency across the tech sector, with platforms doing relatively well.

Table 5.11 Ranking 'algorithmic transparency', Big Tech Scorecard 2022 by Ranking Digital Rights (2022a)

Rank (out of 14)	Platform	Score
1	Meta	22%
4	Twitter	20%
6	YouTube (Google)	14%

Table 5.12 Reporting of content policy-based moderation data (2016–2022)

Platform	2016	2017	2018	2019	2020	2021	2022
Facebook	No	Yes	Yes	Yes	Yes	Yes	Yes
Instagram	No	No	No	Yes	Yes	Yes	Yes
Twitter	No	No	Yes	Yes	Yes	Yes	Yes
TikTok	No	No	No	Yes	Yes	Yes	Yes
YouTube	No	Yes	Yes	Yes	Yes	Yes	Yes

TikTok is not included in the ranking. There are various other projects examining platform moderation transparency. New America's Transparency Report Tracking Tool is a continuously updated project that curates data from transparency reports of six services of five platform companies (Singh and Doty 2021). The tracking tool allows readers to find in one place the categories of transparency reporting included in transparency reports of Facebook, Instagram, Reddit, TikTok, Twitter and YouTube. The tracking tool also allows an over-time view of when certain reporting categories have been added or dropped by the services. What is not included is any attempt to find common categories of transparency reporting that would allow to compare changes over time between the different platforms. On a general level, it is worth examining when the five platforms' services have started to disclose transparency reports concerning moderation actions based on their content policies (as opposed to government requests, etc.). Such longitudinal data is presented in Table 5.12.[4]

As Table 5.12 shows, with regard to content moderation action based on conflict with platform content policies, there appears to be a degree of isomorphism across platforms. The evolution of this common practice is interesting, though. Content policy is actually not one of the first

[4] The underlying data is derived from Quintais et al. (2022).

categories of content removal that was introduced into transparency reporting (Quintais et al. 2022). The amount of content moderated was first shared by Facebook and YouTube in 2017. Only subsequently in 2018, Twitter started to disclose the data for content removed due to its Twitter Rules. Instagram and TikTok started to reveal the data for such platform-policy-based moderation of content in 2019. However, the quality of reporting also matters greatly. On the one hand, what is crucially lacking is a common standard by which data is reported, even if the reporting slowly converges towards common criteria. It remains difficult to make data actually comparable. On the other hand, the protection of human rights requires an in-depth understanding by the public and by policymakers regarding the processes at play, especially concerning the harms that platforms may have data on.

Whether such data is relevant to content moderation can often only be seen once additional reasons to restrict specific content are established. The Facebook Files relate to a recent whistleblowing and succeeding scandal, in which one shortcoming of Meta received a particularly high degree of media attention. The leaks demonstrated that the company had long known the impact of the use of its platforms on the mental-health of young adults, specifically that "Instagram is harmful for a sizable percentage [young users], most notably teenage girls" (Wells et al. 2021). Not being transparent where internal data suggests major issues is highly problematic. For instance, Leightley et al. (2022) argue that access to platform data could be used to better understand the mental-health implications, suggesting that "limited data access by researchers is preventing such advances from being made". Whether content governance would be a tool to tackle these challenges would have to be established. This demonstrates that transparency is required in a serious and comprehensive way in order to protect human rights, rather than being a mere exercise of counting and publishing high-level data.

Overall, this chapter demonstrates a number of noteworthy trends when it comes to integrating human rights claims into platform policies and with regard to both substantial outcomes and procedural content moderation practices. It becomes apparent that the platforms are using the language of human rights but often not in their content policies. With regard to the latter, it can be said that—as far as reported—content moderation outcomes can be better understood through the lenses of the Internet Bills of Rights introduced in Chap. 4. Such a perspective also allows us to observe that the five studied platforms appear to largely converge on a number of indicators. Convergence on a common standard is also a useful narrative to understand practices related to automated content moderation and transparency reporting, even if—particularly with regard to automation of content moderation—there are substantial deviations from civil society demands. Importantly, even a standard of practices on which platforms converge, while referencing human rights at least in name, does not suffice to fully solve the content governance dilemma platforms face. Deep engagement with human rights standards and continuous exchange with those who defend them are needed to ensure human rights are indeed realised. The ongoing process by UNESCO for "Guidelines for Regulating Digital Platforms" (2022) might be an additional way forward, as may be general moves towards so-called platform councils that bring together different stakeholders to counsel platform policy teams (Tworek 2019).

References

AccessNow. 2020. *Open Letter to Facebook, Twitter, and YouTube: Stop Silencing Critical Voices from the Middle East and North Africa, December 2020.* Open Letter. https://www.accessnow.org/facebook-twitter-youtube-stop-silencing-critical-voices-mena/. Accessed December 21, 2022.

Al Ghussain, Alia. 2022. *Meta's Human Rights Report Ignores the Real Threat the Company Poses to Human Rights Worldwide.* Amnesty International. https://www.amnesty.org/en/latest/campaigns/2022/07/metas-human-rights-report-ignores-the-real-threat-the-company-poses-to-human-rights-worldwide/. Accessed February 20, 2023.

Allan, Richard. 2018. Hard Questions: Where Do We Draw the Line on Free Expression? *about.fb.com.* https://about.fb.com/news/2018/08/hard-questions-free-expression/. Accessed September 24, 2022.

Amnesty International. 2022. Myanmar: Facebook's Systems Promoted Violence Against Rohingya; Meta Owes Reparations. https://www.amnesty.org/en/latest/news/2022/09/myanmar-facebooks-systems-promoted-violence-against-rohingya-meta-owes-reparations-new-report/. Accessed February 20, 2023.

Barlow, John Perry. 1996. A Declaration of Independence of Cyberspace. *eff.org.* https://www.eff.org/cyberspace-independence. Accessed October 28, 2022.

Barrett, Bridget, and Daniel Kreiss. 2019. Platform Transience: Changes in Facebook's Policies, Procedures, and Affordances in Global Electoral Politics. *Internet Policy Review* 8 (4): 1–22.

Binns, Reubes, Michael Veale, Max Van Kleek, and Nigel Shadbolt. 2017. Like Trainer, Like Bot? Inheritance of Bias in Algorithmic Content Moderation. In

Social Informatics 9th International Conference, SocInfo 2017, Oxford, UK, September 13–15, 2017, Proceedings, Part II, ed. Giovanni Luca Ciampaglia, Afra Mashhadi, and Taha Yasseri, 405–415. Oxford: Springer.

Bygrave, Lee A. 2015. *Internet Governance by Contract*. Oxford: Oxford University Press.

Celeste, Edoardo. 2019. Terms of Service and Bills of Rights: New Mechanisms of Constitutionalisation in the Social Media Environment? *International Review of Law, Computers & Technology* 33 (2): 122–138. https://doi.org/10.108 0/13600869.2018.1475898.

———. 2022. *Digital Constitutionalism: The Role of Internet Bills of Rights*. London: Routledge.

Citron, Danielle Keats, and Benjamin Wittes. 2017. The Internet Will Not Break: Denying Bad Samaritans Sec. 230 Immunity. *Fordham Law Review* 86 (2).

Conger, Kate. 2018. Google Removes 'Don't Be Evil' Clause from Its Code of Conduct. *gizmodo.com*. https://gizmodo.com/google-removes-nearly-all-mentions-of-dont-be-evil-from-1826153393. Accessed September 22, 2022.

Crocker, Andrew, Gennie Gebhart, Aaron Mackey, Kurt Opsahl, Hayley Tsukayama, Jamie Lee Williams, and Jillian C. York. 2019. Who Has Your Back? *eff.org*. https://www.eff.org/wp/who-has-your-back-2019. Accessed October 30, 2022.

Deutsche Welle. 2015. European Court of Human Rights Rules Turkey's Ban on YouTube Violated Rights. https://www.dw.com/en/european-court-of-human-rights-rules-turkeys-ban-on-youtube-violated-rights/a-18886693. Accessed December 21, 2022.

European Union. 2019. Regulation (EU) 2019/1150. 20 June 2019. *Promoting Fairness and Transparency for Business Users of Online Intermediation Services*. https://eur-lex.europa.eu/eli/reg/2019/1150/oj.

Facebook. 2007. Facebook Community Standards (Version of 30 August 2007). Platform Governance Archive. https://github.com/PlatformGovernanceArchive/pga-corpus/tree/main/Versions/PDF/Facebook. Accessed June 16, 2023.

———. 2021. Facebook Community Standards (Version of 30 January 2021). Platform Governance Archive. https://github.com/PlatformGovernanceArchive/pga-corpus/tree/main/Versions/PDF/Facebook. Accessed June 16, 2023.

Google. 2022. Human rights. *Google.com*. https://about.google/human-rights/. Accessed September 24, 2022.

Gorwa, Robert, Reuben Binns, and Christian Katzenbach. 2020. Algorithmic Content Moderation: Technical and Political Challenges in the Automation of Platform Governance. *Big Data & Society* 7 (1).

Guo, Eileen. 2021. Deplatforming Trump Will Work, Even if it Won't Solve Everything. *MIT Technology Review*. https://www.technologyreview.

com/2021/01/08/1015956/twitter-bans-trump-deplatforming/. Accessed October 20, 2022.

Ha, Anthony. 2021. YouTube Will Start Penalizing Channels that Post Election Misinformation. *TechCrunch*. https://techcrunch.com/2021/01/07/youtube-election-strikes/. Accessed January 20, 2023.

Hemphill, T. A. 2019. 'Techlash', responsible innovation, and the self-regulatory organization. *Journal of Responsible Innovation* 6 (2), 240–247.

Helfer, Laurence R., and Molly K. Land. 2022. The Facebook Oversight Board's Human Rights Future. *Duke Law School Public Law & Legal Theory and Research Paper Series No. 2022-47* 44 (6). https://doi.org/10.2139/ssrn.4197107.

Horwitz, Jeff. 2021. Facebook Says Its Rules Apply to All. Company Documents Reveal a Secret Elite That's Exempt. *The Wall Street Journal*. https://www.wsj.com/articles/facebook-files-xcheck-zuckerberg-elite-rules-11631541353. Accessed December 21, 2022.

Iqbal, Mansoor. 2022. TikTok Revenue and Usage Statistics (2022). *Businessofapps.com*. https://www.businessofapps.com/data/tik-tok-statistics/. Accessed September 24, 2022.

Katzenbach, Christian. 2021. AI Will Fix This—The Technical, Discursive, and Political Turn to AI in Governing Communication. *Big Data & Society* 8 (2). https://doi.org/10.1177/20539517211046182.

Kelly, Makena. 2021. Biden Revokes and Replaces Trump Orders Banning TikTok and WeChat. *The Verge*. https://www.theverge.com/2021/6/9/22525953/biden-tiktok-wechat-trump-bans-revoked-alipay. Accessed September 24, 2022.

Kettemann, Matthias, and Wolfgang Schulz. 2020. Setting Rules for 2.7 Billion. A (First) Look into Facebook's Norm-Making System: Results of a Pilot Study. *Working Papers of the Hans-Bredow-Institut, Works in Progress# 1*. Hamburg: Hans-Bredow-Institut.

Klonick, K. 2018. The New Governors: The People, Rules, and Processes Governing Online Speech. *Harvard Law Review* 131 (6): 1598–1670.

Kuczerawy, Aleksandra, and Jef Ausloos. 2015. From Notice-and-Takedown to Notice-and-Delist: Implementing Google Spain. *Colorado Technology Law Journal* 14 (2): 219–258.

Leightley, Daniel, Amanda Bye, Ben Carter, Kylee Trevillion, Stella Branthonne-Foster, Maria Liakata, Anthony Wood, Dennis Ougrin, Amy Orben, Tamsin Ford, and Rina Dutta. 2022. Maximising the Positive and Minimising the Negative: Social Media Data to Study Youth Mental Health with Informed Consent. *Frontiers in Psychiatry* 13.

Lumen. 2022. About us. *lumendatabase.org*. https://lumendatabase.org/pages/about. Accessed October 28, 2022.

Meta. 2022a. Facebook Community Standards. *transparency.fb.com*. https://transparency.fb.com/policies/community-standards/. Accessed December 21, 2022.

———. 2022b. Community Standards Enforcement Report. *transparency.fb.com*. https://transparency.fb.com/data/community-standards-enforcement/. Accessed December 21, 2022.

———. 2022c. Corporate Human Rights Policy. https://about.fb.com/wp-content/uploads/2021/03/Facebooks-Corporate-Human-Rights-Policy.pdf. Accessed December 14, 2022.

Milmo, Dan. 2022. Twitter Says it Suspends 1m Spam Users a Day as Elon Musk Row Deepens. https://www.theguardian.com/technology/2022/jul/07/twitter-says-it-suspends-1m-spam-users-a-day-as-elon-musk-dispute-deepens. Accessed September 22, 2022.

Nonnecke, Brandie, and Camille Carlton. 2022. EU and US Legislation Seek to Open up Digital Platform Data. *Science* 375 (6581): 610–612.

Oversight Board. 2020. Breast Cancer Symptoms and Nudity. Case 2020-004-IG-UA. https://www.oversightboard.com/decision/IG-7THR3SI1

———. 2021. Former President Trump's Suspension. Case 2021-001-FB-FBR. https://www.oversightboard.com/decision/FB-691QAMHJ/

———. 2022. Iran Protest Slogan. Case 2022-013-FB-UA. https://www.oversightboard.com/decision/FB-ZT6AJS4X/

Quintais, João Pedro, Péter Mezei, István Harkai, João Carlos Magalhaes, Christian Katzenbach, Sebastian Felix Schwemer, and Thomas Riis. 2022. *Copyright Content Moderation in the EU: An Interdisciplinary Mapping Analysis*. reCreating Europe Report. https://papers.ssrn.com/sol3/papers.cfm?abstract_id=4210278.

Ranking Digital Tech. 2022a. The Big Tech Scorecard 2022. *Rankingdigitalrights.org*. https://rankingdigitalrights.org/index2022/. Accessed October 31, 2022.

———. 2022b. Key Findings from the 2022 RDR Big Tech Scorecard. *Rankingdigitalrights.org*. https://rankingdigitalrights.org/bts22/. Accessed October 5, 2022.

Singh, Spandana, and Doty, Leila. 2021. The Transparency Report Tracking Tool: How Internet Platforms Are Reporting on the Enforcement of Their Content Rules. *Newamerica.org*. https://www.newamerica.org/oti/reports/transparency-report-tracking-tool/. Accessed October 28, 2022.

Statista. 2022. Most Popular Social Networks Worldwide as of January 2022, Ranked by Number of Monthly Active Users. *statista.com*. https://www.statista.com/statistics/272014/global-social-networks-ranked-by-number-of-users/. Accessed September 23, 2022.

Suzor, Nicolas. 2019. *Lawless: The Secret Rules that Govern Our Digital Lives*. Cambridge: Cambridge University Press.

Suzor, Nicolas, Tess Van Geelen, and Sarah Myers West. 2018. Evaluating the Legitimacy of Platform Governance: A Review of Research and a Shared Research Agenda. *International Communication Gazette* 80 (4): 385–400. https://doi.org/10.1177/1748048518757142.

Tewari, Shreya. 2022. *Transparency Initiatives in the DSA: An Exciting Step Forward in Transparency Reporting. Lumen Project.* https://www.lumendatabase.org/blog_entries/transparency-initiatives-in-the-dsa-an-exciting-step-forward-in-transparency-reporting. Accessed October 28, 2022.

The Guardian. 2022. TikTok Moves to Ease Fears Amid Report Workers in China Accessed US Users' Data. https://www.theguardian.com/technology/2022/jun/17/tiktok-us-user-data-china-bytedance. Accessed September 22, 2022.

TikTok. 2021a. Community Guidelines Enforcement Report, January 1, 2021–March 31, 2021. *tiktok.com.* https://www.tiktok.com/transparency/en-us/community-guidelines-enforcement-2021-1/. Accessed January 12, 2023.

———. 2021b. Community Guidelines Enforcement Report, April 1, 2021–June 30, 2021. *tiktok.com.* https://www.tiktok.com/transparency/en-us/community-guidelines-enforcement-2021-2/. Accessed January 12, 2023.

———. 2022a. Upholding Human Rights. *tiktok.com.* https://www.tiktok.com/transparency/en-au/upholding-human-rights/. Accessed November 24, 2022.

———. 2022b. Community Guidelines. *tiktok.com.* https://www.tiktok.com/community-guidelines?lang=en. Accessed November 24, 2022.

———. 2022c. Community Guidelines Enforcement Report, July 1, 2021–September 30, 2021. *tiktok.com.* https://www.tiktok.com/transparency/en-us/community-guidelines-enforcement-2021-3/. Accessed January 12, 2023.

———. 2022d. Community Guidelines Enforcement Report, October 1, 2021–December 31, 2021. *tiktok.com.* https://www.tiktok.com/transparency/en-us/community-guidelines-enforcement-2021-4/. Accessed January 12, 2023.

———. 2022e. Community Guidelines Enforcement Report, July 1, 2022–September 30, 2022. *tiktok.com.* https://www.tiktok.com/transparency/en-us/community-guidelines-enforcement-2022-3/. Accessed January 13, 2023.

Twitter. 2009. The Twitter Rules (Version of 18 January 2009). Platform Governance Archive. https://github.com/PlatformGovernanceArchive/pga-corpus/tree/main/Versions/PDF/Twitter. Accessed June 16, 2023.

———. 2020. The Twitter Rules (Version of 28 October 2020). Platform Governance Archive. https://github.com/PlatformGovernanceArchive/pga-corpus/tree/main/Versions/PDF/Twitter. Accessed June 16, 2023.

———. 2021. The Twitter Rules (Version of 27 January 2021). Platform Governance Archive. https://github.com/PlatformGovernanceArchive/pga-corpus/tree/main/Versions/PDF/Twitter. Accessed June 16, 2023.

———. 2022a. Defending and Respecting the Rights of People Using Our Service. *twitter.com.* https://help.twitter.com/en/rules-and-policies/defending-and-respecting-our-users-voice. Accessed September 18, 2022.

———. 2022b. Rules Enforcement. *Transparency.twitter.com*. https://transparency.twitter.com/en/reports/rules-enforcement.html.   Accessed January 13, 2023.

Tworek, Heidi. 2019. Social Media Councils. In *Models for Platform Governance—A CIGI Essay Series*, 97–102. Waterloo: Centre for International Governance Innovation.

UNESCO. 2022. *Guidelines for Regulating Digital Platforms: A Multistakeholder Approach to Safeguarding Freedom of Expression and Access to Information*. CI-FEJ/FOEO/3   Rev.   https://unesdoc.unesco.org/ark:/48223/pf0000384031

United Nations. 2018. *Report of the Special Rapporteur on the Promotion and Protection of the Right to Freedom of Opinion and Expression, David Kaye*. UN Doc A/73/348.

———. 2019. *Report of Report of the Special Rapporteur on the Promotion and Protection of the Right to Freedom of Opinion and Expression, David Kaye*. UN Doc A/47/486.

Wells, Georgia, Jeff Horwitz, and Deepa Seetharaman. 2021. Facebook Knows Instagram Is Toxic for Teen Girls, Company Documents Show. *The Wall Street Journal*. https://www.wsj.com/articles/facebook-knows-instagram-is-toxic-for-teen-girls-company-documents-show-11631620739?mod=hp_lead_pos7&mod=article_inline. Accessed September 22, 2022.

YouTube. 2022a. Community Guidelines. *youtube.com*. https://www.youtube.com/howyoutubeworks/policies/community-guidelines/.   Accessed September 24, 2022.

———. 2022b. Transparency Report. *youtube.com*. https://transparencyreport.google.com/youtube-policy/. Accessed January 14, 2022.

Zuboff, Shoshana. 2019. *The Age of Surveillance Capitalism: The Fight for a Human Future at the New Frontier of Power*. London: Profile Books.

Zuckerberg, Mark. 2019. *Facebook's Commitment to the Oversight Board, September 2019*. Open Letter. https://about.fb.com/wp-content/uploads/2019/09/letter-from-mark-zuckerberg-on-oversight-board-charter.pdf.   Accessed September 24, 2022.

CHAPTER 6

Conclusion

Abstract The choice on the law of online content moderation is a conundrum composed of three options, none of which is fully satisfying. International human rights law is not *the* solution to the content governance dilemma, but its contribution cannot be excluded from the resolution of this issue. A multi-level approach is proposed where multiple actors are instrumental in translating human rights norms into rules that speak to the context of online content governance. This work focuses on the role of civil society organisations. By comparing the demands advanced by civil society actors with platforms' policies we reconstruct an image of the process of social media constitutionalisation 'in motion'.

Keywords Online content moderation • Social media • Civil society • Constitutionalisation • Multi-level approach • Digital constitutionalism

Online content moderation is affected by multiple human rights concerns. This book has mentioned some of them, from the use of discriminatory and opaque algorithms to the lack of procedural rules. Yet, this work has focused on a theme that is common to all of them. We have analysed a meta-problem: the transversal issue of clarifying which rules should govern online content moderation, and thus help it prevent the aforementioned human rights concerns. We have done it without taking a normative

© The Author(s) 2023
E. Celeste et al., *The Content Governance Dilemma*, Information
Technology and Global Governance,
https://doi.org/10.1007/978-3-031-32924-1_6

approach, striving to present an objective account of the discrepancies between legal theory and reality.

Firstly, by exposing the dilemmatic nature of the choice on the law of online content moderation (see Chap. 2). A conundrum composed of three options, none of which is fully satisfying. If social media platforms adopt content policies based on their own values—for example, Facebook's 'voice, authenticity, safety, privacy and dignity'—they are unavoidably accused of unilaterally setting the rules of the game for their own virtual spaces, questioning why they depart from international human rights law and suspecting that these values are a way to protect their business interest. However, if this form of platform authoritarianism is to exclude, one cannot knowingly affirm that the solution is merely to refer to existing laws. Social media are global spaces characterised by a melting pot of users coming from various countries: Which national law could or should prevail without being accused of digital imperialism? Moreover, resorting to international law standards is not the panacea that one would expect, but would rather lead to a situation of normative anomie, a status of disorientation justified by the fact that its norms do not directly target private actors, and only include very general principles that would in any case, require a further interpretation in order to be applied in the context of online content moderation.

Yet, if this book shows how international human rights law is not *the* solution to the content governance dilemma, it does not exclude its contribution to the resolution of this issue (see Chap. 3). We propose a multilevel approach where multiple actors are instrumental in translating human rights norms—what constitutes the DNA of contemporary constitutionalism—into rules that speak to the context of online content governance. Constitutionalism is the ideology that champions the respect of fundamental rights through the limitation of the power of dominant actors. It is embedded into international human rights law, but regrettably these norms speak to a social reality that has been overtaken. Today, the multinational companies owning and managing social media platforms emerge as dominant actors beside nation states. Achieving a form of digital constitutionalism would consist in rearticulating its principles in the context of the digital society, subjecting social media platforms to the same types of obligations that international law imposes on states. Constitutionalising social media would therefore mean instilling human rights guarantees in an environment that very often lacks them due to its conceptualisation as a private space (Celeste et al. 2022).

In this work we focus on an actor that is often neglected, particularly among legal scholars: civil society organisations. From a legal point of view, as strong the voice of these actors might be, their claims remain outside what is considered to be legally relevant as their normative power does not have legal force. Yet, this book shows how civil society, if it is on the one hand unable to produce *lex*—something that social media platforms as owners of their virtual fiefdoms can conversely do—can and does contribute to the development of the *ius*, the legal discourse, on the rules that should govern online content moderation. Adopting a musical metaphor, our work has consisted in putting a series of fragmented voices together in one score, so that it can be read—or played—in unison. The music that one can read from this analysis is a vocal appeal to clarify what the legitimate limitations to the principle of freedom of expression that social media platforms should apply online are; to establish procedural principles in order to mitigate the potential arbitrariness of social media platforms' decisions. Civil society actors thus 'generalise and recontextualise' principles of international human rights law and core constitutional values in a way that directly addresses the challenges of the social media environment (Celeste 2022; Teubner 2012). Not only in the form of meta-rules but also by directly embedding norms into the socio-technical architecture that run and govern online platforms: a form of constitutionalism 'by design' and 'by default' which holds together technical solutions and governance mechanisms (see Chap. 4).

By comparing the demands advanced by civil society actors with platforms' policies we have reconstructed an image of this process of social media constitutionalisation 'in motion' (see Chap. 5). Over the past few years, there has indeed been a positive trend towards an increased proceduralisation and transparency of online content moderation. There is a progressive convergence between civil society demands and platforms' policies. The use of automated systems to filter and take down content is now compensated by increasing numbers of regulated appeal mechanisms—at times even multiple, as in the case of the two-instance structure created by Meta with the creation of the Oversight Board. Most social media companies publish detailed transparency reports that help reduce the level of opacity characterising the governance of content moderation and increase accountability towards the general public, researchers and public authorities.

Certainly, this is not the outcome of civil society advocacy alone. There are dramatic moments—as always in history—from which social media companies are learning, like after the assault on Capitol Hill. There are still tensions between a proprietary vision of social media as private fiefdoms, whose internal rules can be modified with a simple tweet by an almighty CEO, and their role as public forum, a contemporary centre for the exercise of fundamental freedoms. There is a growing body of law, both at national and at regional level, directly addressing content moderation practices in order to tackle the issue of online harm. Courts and internal oversight bodies are playing a 'maieutic' role in interpreting and further developing platforms' policies in a fundamental rights-compliant way (Celeste 2021; Karavas 2010). There is now a shared belief that social media companies can no longer be left entirely alone in regulating online content. Thanks to contributions from various actors, the process of constitutionalisation of social media is shaping clearer rules guaranteeing digital human rights.

The solution to the dilemma on which rules should govern online content moderation can thus be found in the structure of the dilemma itself. It is its composite nature, the key to understanding how to legitimately develop norms that will be able to preserve fundamental rights on social media platforms. It is not a question of choosing which actor should prevail and impose its law. The solution rather lies in recomposing the puzzle of the various voices that are contributing to shape digital human rights in the context of online platforms.

References

Celeste, E. 2021. Digital Punishment: Social Media Exclusion and the Constitutionalising Role of National Courts. *International Review of Law, Computers & Technology* 35: 162–184. https://doi.org/10.1080/1360086 9.2021.1885106.

———. 2022. *Digital Constitutionalism: The Role of Internet Bills of Rights.* Routledge.

Celeste, E., A. Heldt, and C. Iglesias Keller, eds. 2022. *Constitutionalising Social Media.* Hart.

Karavas, V. 2010. Governance of Virtual Worlds and the Quest for a Digital Constitution. In *Governance of Digital Game Environments and Cultural Diversity: Transdisciplinary Enquiries,* ed. C.B. Graber and M. Burri-Nenova, 153–169. Edward Elgar Publishing.

Teubner, G. 2012. *Constitutional Fragments: Societal Constitutionalism and Globalization.* Oxford University Press.

ANNEX: LIST OF ANALYSED DOCUMENTS

Access Now. 2020. 26 Recommendations on Content Governance. https://www.accessnow.org/cms/assets/uploads/2020/03/Recommendations-On-Content-Governance-digital.pdf.

ACLU Foundation et al. 2018. The Santa Clara Principles on Transparency and Accountability in Content Moderation. https://santaclaraprinciples.org.

African Declaration Coalition, African Declaration on Internet Rights and Freedoms. 2014. http://africaninternetrights.org/articles/.

Aspen Institute. 2011. The Aspen IDEA Principles. http://www.umic.pt/images/stories/publicacoes5/Aspen%20IDEA%20Project.pdf.

Association for Progressive Communications. 2006. Internet Rights Charter. https://www.apc.org/node/5677.

Association for Progressive Communications. 2011. Statement Internet rights are human rights. https://www.apc.org/en/pubs/briefs/internet-rights-are-human-rights-claims-apc-human-.

Association for Progressive Communications. 2018. Content Regulation in the Digital Age. https://www.ohchr.org/Documents/Issues/Opinion/ContentRegulation/APC.pdf.

Gender, Sexuality, and Internet Meeting organised by Association for Progressive Communications. 2014. Feminist Principles of the Internet. https://feministinternet.org/sites/default/files/Feminist_principles_of_the_internetv2-0.pdf.

© The Author(s) 2023
E. Celeste et al., *The Content Governance Dilemma*, Information Technology and Global Governance,
https://doi.org/10.1007/978-3-031-32924-1

Article 19. 2017. Universal Declaration of Digital Rights. https://www.article19.org/resources/internetofrights-creating-the-universal-declaration-of-digital-rights/.

Article 19. 2017. Getting connected: Freedom of expression, telcos and ISPs. https://www.article19.org/wp-content/uploads/2017/06/Final-Getting-Connected-2.pdf.

Article 19. 2018. Self-regulation and 'hate speech' on social media platforms. http://europeanjournalists.org/mediaagainsthate/wp-content/uploads/2018/02/Self-regulation-and-%E2%80%98hate-speech%E2%80%99-on-social-media-platforms_final_digital.pdf.

Cambodian Center for Independent Media. 2015. Statement of Principles for Cambodian Internet Freedom. http://www.cchrcambodia.org/media/files/press_release/573_201jsopfcife_en.pdf.

Contract for the Web. 2018. https://contractfortheweb.org.

Chinese Academics. 2009. China Internet Human Rights Declaration. https://hawkeyejack.blogspot.com/2009/10/chinas-call-for-internet-freedom.html?m=0.

Civil Society—TUAC. 2008. Seul Declaration. https://thepublicvoice.org/PVevents/seoul08/seoul-declaration.pdf.

Civil Society (WSIS). 2003. Civil Society Declaration to the World Summit on the Information Society. https://www.itu.int/net/wsis/docs/geneva/civil-society-declaration.pdf.

Civil Society. 2017. Declaration on Freedom of Expression In response to the adoption of the Network Enforcement Law ("Netzwerkdurchsetzungsgesetz") by the Federal Cabinet on April 5, 2017. https://deklaration-fuer-meinungsfreiheit.de/en/.

EDRi. 2014. The Charter of Digital Rights.https://edri.org/wp-content/uploads/2014/06/EDRi_DigitalRightsCharter_web.pdf.

Robert B. Gelman. 1997. Declaration of Human Rights in Cyberspace. http://www.be-in.com/10/rightsdec.html.

Global Forum for Media Development. 2019. GFMD Statement on the Christchurch Call and Countering Violent Extremism Online. https://drive.google.com/file/d/1N4EwiM7eITD6plQrYqJI02NayvCiMCT-/view.

Global Network Initiative. 2008. Principles on Freedom of Expression and Privacy. https://www.globalnetworkinitiative.org/principles/index.php.

iRights Coalition. 2015. iRights. http://irights.uk/the_irights/.

Internet Rights and Principles Coalition. 2010. The Charter of Human Rights and Principles for the Internet. http://internetrightsandprinciples.org/site/charter/.

Internet Rights and Principles Coalition- Internet Governance Forum. 2014. The Charter of Human Rights and Principles for the Internet 2.0 4th ed. http://www.ohchr.org/Documents/Issues/Opinion/Communications/InternetPrinciplesAndRightsCoalition.pdf.

Jeff Jarvis. 2010. A Bill of Rights in Cyberspace. http://buzzmachine.com/2010/03/27/a-bill-of-rights-in-cyberspace/.

Just Net Coalition. 2014. Delhi Declaration for a Just and Equitable Internet. http://justnetcoalition.org/delhi-declaration.

Andrew Murray. 2010. A Bill of Rights for the Internet. http://theitlawyer.blogspot.com/2010/10/bill-of-rights-for-internet.html.

NetMundial. 2014. NetMundial Statement. http://netmundial.br/wp-content/uploads/2014/04/NETmundial-Multistakeholder-Document.pdf.

New Zealand Civil Society. 2019. Community Input on Christchurch Call. https://docs.google.com/document/d/10RadyVQUNu1H5D7x6IJVKbqmaeDeXre0Mk-FFNkIVxs/edit?ts=5cda8902#.

New Zealand Green Party. 2014. Internet Rights and Freedoms Bill. https://home.greens.org.nz/misc-documents/internet-rights-and-freedoms-bill.

People's Communication Charter Coalition. 1999. People's Communication Charter. http://www.pccharter.net/charteren.html.

Praxis Centre for Policy Studies. 2012. Guiding Principles of Internet Freedom. http://www.praxis.ee/fileadmin/tarmo/Projektid/Valitsemine_ja_kodanike%C3%BChiskond/Praxis_Theses_Internet.pdf.

PEN International. 2011. Declaration on Digital Freedom. http://www.pen-international.org/declaration-on-digital-freedom-english/.

Parliamentary Human Rights Foundation (PHRF) e Open Society Institute. 1997. Open Internet Policy Principles. http://mailman.anu.edu.au/pipermail/link/1997-March/026302.html.

Free Internet Activism Sub-group—Reddit. 2012. The Digital Bill of Rights—A Crowd-sourced Declaration of Rights. https://www.reddit.com/r/fia/comments/vuj37/digital_bill_of_rights_1st_draft/.

Reporters Sans Frontiers. 2018. International Declaration on Information and Democracy. https://rsf.org/en/global-communication-and-information-space-common-good-humankind.

Tech Freedom et al. 2012. Declaration of Internet Freedom. http://declarationofinternetfreedom.org/.

UK Liberal Democrats. 2015. Protecting your data online with a Digital Rights Bill. http://www.libdems.org.uk/protecting-your-data-online-with-a-digital-rights-bill.

World Federation of Scientists. 2009. Erice Declaration on Principles for Cyber Stability and Cyber Peace. http://www.aps.org/units/fip/newsletters/201109/barletta.cfm.

Zeit-Foundation. 2016. Charter of Digital Fundamental Rights of the European Union. https://digitalcharta.eu/wp-content/uploads/2016/12/Digital-Charta-EN.pdf.

Index[1]

---

[1] Note: Page numbers followed by 'n' refer to notes.